CELL BIOLOGY RESEARCH PROGRESS

FLOW CYTOMETRY

PRINCIPLES, METHODOLOGY AND APPLICATIONS

Cell Biology Research Progress

Additional books in this series can be found on Nova's website under the Series tab.

Additional e-books in this series can be found on Nova's website under the e-book tab.

CELL BIOLOGY RESEARCH PROGRESS

FLOW CYTOMETRY

PRINCIPLES, METHODOLOGY AND APPLICATIONS

STEFANOS PAPANDREOU
EDITOR

NOVA BIOMEDICAL
New York

For permission to use material from this book please contact us:
Telephone 631-231-7269; Fax 631-231-8175
Web Site: http://www.novapublishers.com

Additional color graphics may be available in the e-book version of this book.

Library of Congress Cataloging-in-Publication Data

Flow cytometry : principles, methodology and applications / editor, Stefanos Papandreou.
pages cm
Includes index.
ISBN: 978-1-62808-709-3 (hardcover)
1. Flow cytometry. I. Papandreou, Stefanos.
QH585.5.F56F5866 2013
616.07'582--dc23

2013029202

Published by Nova Science Publishers, Inc. † New York

Contents

Preface

Flow cytometry is a method to conduct a multiparameter analysis of cells suspended in liquid and passing through a laser beam. As cells pass through the flow chamber they are hit with the laser light beam which is scattered in different directions and recorded as forward light scatter and side light scatter. In this book, the authors discuss the principles, methodology and applications of flow cytometry. Topics include flow cytometry and epifluorescence analyses of freshwater bacterio- and virioplankton communities; application of flow cytometry in phenotyping, analysis and functional characterization of T regulatory and immune suppressor cells; multicolor panel design strategies in rare event analysis of dendritic cells and regulatory T cells; advanced characterization of individual particle morphology light-scattering flow cytometry; and application of flow cytometry in environmental science.

Chapter 1 – Concerns about obtaining accurate determinations of the concentrations of viruses and bacteria in freshwater samples led us to examine a broad battery of counting and storage procedures for use in flow cytometry (FCM) and epifluorescence microscopy (EFM) analyses. Sample preparations were done so as to optimize counts and preservation by using different types and concentrations of aldehyde-based fixatives, stains belonging to the SYBR family, dilution media, temperature and storage conditions. Whenever possible, FCM and EFM counts were compared. Results obtained using FCM supported the addition of fixative for bacteria, preferably glutaraldehyde at a final concentration of 2%, dilution in 0.2-µm or 0.02-µm filtered Tris-EDTA buffer (TE, pH = 7.5), staining with SYBR Green I at a final concentration of 10^{-4} and incubating at ambient temperature for at least 15 minutes. For viruses, there was no need to add fixative, whereas dilution in recently-autoclaved and 0.02-µm filtered TE and incubation with SYBR Gold at a final concentration of 2×10^{-5} at 75°C for 10 minutes is recommended. If possible, FCM samples should be counted on day = 0, although the authors do show that bacterioplankton samples, at least, may be stored at 4°C and counted at 24 h later but not beyond if samples cannot be frozen in liquid nitrogen. The conditions required for optimum EFM counts of both bacteria and viruses involved were to stain filters with SYBR Gold at a final concentration of 10^{-3}. Slides could be counted for up to 1 month if rapidly frozen and stored at –20°C.

The authors results performed on lake samples clearly demonstrate the importance of defining the best conditions in order to get reliable counts of microbial communities such as viruses and bacteria. Each research laboratory should undertake such tests according to the equipment available, and the needs and area of their research.

Chapter 2 – Flow cytometry is an inevitable tool in clinical diagnosis of various pathological conditions. Recently, its application plays a significant role in monitoring of several immune suppressor cells in patients with cancer and autoimmune disorders. Regulatory T cells (Treg cells) are immune cells which help in maintaining immune homeostasis in physiological conditions by patrolling unwanted immune responses; however, in cancers and autoimmune diseases, Treg cell numbers and function are significantly altered. In this chapter, the authors focus and update the information regarding Treg cells and their subsets, and application of flow cytometry in phenotyping and functional characterization of various immune regulatory and suppressor cells in multiple myeloma patients.

Chapter 3 – The need for multidimensional analysis using complex stain combinations has been expanding considerably, especially to discriminate, classify and/or isolate rare cell populations, such as dendritic cells (DCs) and regulatory T cells (Treg). Several principles need to be taken into account when setting up a multicolor flow cytometry experiment for rare event analysis.

The frequency of the event of interest in a sample dictates how many cells must be processed. The lower the frequency of the events of interest, the more events need to be acquired, which might result in large data files. However, the more events one counts, the better the precision will be, as the minimal counting error is equal to the square root of the actual counted events of interest.

It is crucial to understand that reliably counting a rare event is like finding a needle in a haystack. Moreover, the spurious events (``noise``) need to be recognized as well, in order to exclude them, determine what causes them and minimize their occurrence. The signal-to noise ratio can be increased in several ways. Firstly, it is very important to carefully select the reagents to improve the discrimination of different populations in multicolor flow cytometry, which depends on the instrument configuration. Furthermore, the rare event should be positively identified by multiple fluorescence parameters, because the random noise is diluted out by a power function of the number of parameters. Secondly, a critical step in panel design is to match the fluorochromes by brightness according to antigen density. The golden rule is to define the antigens with the lowest expression by the brightest chromophore and vice versa. Thirdly, fluorescence spillover can have a huge impact on the resolution sensitivity of dim markers in multicolor flow cytometry experiments. The more colors you have in your tube, the more spillover you have, resulting in a higher background. Also, if multiple antigens are present on a given cell ("coexpression"), it is better to spread them across as many lasers as possible. Finally, the total noise can be characterized by using an appropriate negative gating control, such as isotype-matched fluorochrome-conjugated antibodies or the fluorescence minus one method.

Here, the authors will discuss all the principles that need to be taken into consideration for a successful multicolor panel design to analyze rare events. They will also illustrate multicolor panel design strategies by specifying the flow cytometric identification and quantification of two rare populations: DC and Treg.

Chapter 4 – Light scattering has played an important role in flow cytometry since the beginning of its technological development. Most of the significant results were demonstrated with analyses of individual particles from light scattering. However the progress in fluorescent biomarkers reduced the role of light scattering in flow-cytometer analyses substantially. Nevertheless the light scattering is still alive and the authors are presenting results from a new stage in the development of light-scattering analysis in flow cytometry.

These new advances stem from angle-resolved methodology. With this review the authors show how the angle-resolved light scattering provides determination of physical characteristics of individual particles. Moreover this approach allows one to retrieve particle dimensions with a precision ranging from 10 nm to 100 nm depending on the shape complexity of a particle. This is subdiffraction precision that cannot be realized with conventional optical microscopy. Moreover the flow-cytometer analysis provides statistically proven results for characterization of cell populations because of the high rate of measurements. In order to demonstrate the advantages of angle-resolved measurement of light scattering the authors describe the principles of scanning flow cytometry that allows us to measure the light-scattering profiles (LSPs) of individual particles. In the development of this approach, they introduce the latest improvement that relates to simultaneous measurement of regular and polarized LSPs for one particle. The methodological goal of the angle-resolved technique is the determination of morphological characteristics of particles from a solution of the inverse light-scattering (ILS) problem. The authors introduce a few methods to solve the ILS problem utilizing the numerical algorithm in global optimization. The bisphere characteristics, sizes, and refractive indices of each sphere composing the bisphere were successfully retrieved by processing the regular and polarized LSPs. The LSPs showed a near-perfect agreement with T-matrix simulations, resulting in a 50-nm precision for sizing bispheres. They rigorously characterized individual blood microparticles, determining their sizes and refractive indices from measured LSPs. The authors introduce a novel approach for determination of the volume and shape of individual blood platelets modeled as oblate spheroids from angle-resolved light scattering using the flow-cytometer technique. Accurate measurement of the spheroidal aspect ratio of the platelets allowed us to discover the substantial differences between native and activated 10 μM adenosine diphosphate samples of platelets. They demonstrate a flow-cytometer method to measure the morphology of single *Escherichia coli* cells with a precision ranging from 50 nm to 250 nm and from 5 nm to 25 nm for bacteria length and diameter, respectively. High precision in the determination of cell morphology allows us to measure kinetics of intracellular processes like red-blood-cell lysis and lymphocyte apoptosis. The authors discuss further applications of angle-resolved light scattering using flow cytometry in the conclusion.

Chapter 5 – Flow cytometry (FCM) is a powerful technology which is routinely used in the medical and veterinary diagnostics. Nowadays it is also an emerging research tool applied in the different fields of environmental studies. This short communication provides the insight to the interesting applications of FCM developed recently in order to study microorganisms, plants and animals. Examples of possible implementation of FCM in environmental monitoring are also given. Major advantages of FCM include automatism and precision of measurements; FCM analyses are time-saving and require small volumes of sample. The authors believe that an increase in FCM availability on the market and simultaneous decrease of technology costs will firmly establish FCM in the different fields of environmental studies and help to improve already existing methods.

In: Flow Cytometry
Editor: Stefanos Papandreou

ISBN: 978-1-62808-709-3

Chapter 1

A Few Tests Prior to Flow Cytometry and Epifluorescence Analyses of Freshwater Bacterio - And Virioplankton Communities

***Stéphan Jacquet*[1,*]*, Ursula Dorigo*[1]* and Sébastien Personnic*[1]
[1]INRA, UMR CARRTEL, Thonon cedex, France

Abstract

Concerns about obtaining accurate determinations of the concentrations of viruses and bacteria in freshwater samples led us to examine a broad battery of counting and storage procedures for use in flow cytometry (FCM) and epifluorescence microscopy (EFM) analyses. Sample preparations were done so as to optimize counts and preservation by using different types and concentrations of aldehyde-based fixatives, stains belonging to the SYBR family, dilution media, temperature and storage conditions. Whenever possible, FCM and EFM counts were compared. Results obtained using FCM supported the addition of fixative for bacteria, preferably glutaraldehyde at a final concentration of 2%, dilution in 0.2-µm or 0.02-µm filtered Tris-EDTA buffer (TE, pH = 7.5), staining with SYBR Green I at a final concentration of 10^{-4} and incubating at ambient temperature for at least 15 minutes. For viruses, there was no need to add fixative, whereas dilution in recently-autoclaved and 0.02-µm filtered TE and incubation with SYBR Gold at a final concentration of 2 x 10^{-5} at 75°C for 10 minutes is recommended. If possible, FCM samples should be counted on day = 0, although we do show that bacterioplankton samples, at least, may be stored at 4°C and counted at 24 h later but not beyond if samples cannot be frozen in liquid nitrogen. The conditions required for optimum EFM counts of both bacteria and viruses involved were to stain filters with SYBR Gold at a final concentration of 10^{-3}. Slides could be counted for up to 1 month if rapidly frozen and stored at –20°C.

Our results performed on lake samples clearly demonstrate the importance of defining the best conditions in order to get reliable counts of microbial communities such

* E-mail: sjacquet@thonon.inra.fr. Tel: +33.4.50.26.78.12; Fax: +33.4.50.26.07.60.

as viruses and bacteria. Each research laboratory should undertake such tests according to the equipment available, and the needs and area of their research.

Keywords: Virus, bacteria, freshwater, enumeration, flow cytometry, epifluorescence microscopy

Introduction

Bacteria and viruses have been shown to be key components of aquatic microbial communities because of their abundance, ubiquity and impact on nutrient cycling, energy fluxes and microbial food webs (Azam et al. 1983; Fuhrman 1999; Wilhelm and Suttle 1999). Bacterioplankton is mainly responsible for the recycling of nutrients, the decomposition of organic matter and for most of the oxygen uptake in the pelagic zone of freshwater ecosystems (Fisher et al. 2000). Bacterioplankton is also a major food source for small and large protozoa (Berninger 1991; Domazion et al. 2003; Simek et al. 1990). Viruses are important in the control of plankton community composition, diversity and succession, and play a key role in bacterioplankton cell mortality (Sime-Ngando et al. 2003; Weinbauer and Rassoulzadegan 2004; Wommack and Colwell 2000, Jacquet et al. 2010) with impacts that vary according to the ecosystem, time and space. They are responsible for 10-60% of the daily bacterioplankton mortality (Bettarel et al. 2003, 2004; Fischer and Velimirov 2002; Jacquet et al. 2005; Simek et al. 2001), and are the most abundant biological particles in both the marine and freshwater environments, with typically 10^7-10^9 viruses.mL^{-1} (Wommack and Colwell 2000; Jacquet et al. 2010). Bacterioplankton densities typically range from 10^5 to 10^7 cells.mL^{-1} (Berthenuis et al. 2012). These densities can vary considerably both with time and space (Larsen et al. 2004; Ovreas et al. 2003; Schröder et al. 2003) due to the influence of physico-chemical and/or biological parameters.

In order to get a better understanding of the ecology of these microorganisms, and to elucidate their role in aquatic systems, we need high frequency water sampling (in both time and space) and subsequent accurate and rapid determinations of their abundances. In the 1970's bacteria were quantified by transmission electron microscopy (TEM) (Watson et al. 1977) or by epifluorescence microscopy (EFM) on acridine stained samples (Francisco et al. 1973). First estimates of viral numbers were obtained using TEM after ultrafiltration (Paul et al. 1991; Proctor and Fuhrman 1990) or ultracentrifugation (Bergh et al. 1989; Bergström and Jansson 2000; Borsheim et al. 1990, Bratbak and Heldal 1993; Sime-Ngando 1997). However, this technique is not only tedious and very time consuming, but also involves some uncertainties arising from the concentration procedures, still requires expensive equipment and skilled personnel. These features make it unsuitable for routine field analysis.

Since the 1990's, the use of EFM in conjunction with the development of new, highly-fluorescent nucleic acid dyes rapidly supplanted TEM, since it was a quicker and less expensive technology (Hara et al. 1991; Hennes and Suttle 1995; Lisle et al. 2004). Nowadays, aquatic bacteria and viruses may be counted by flow cytometry (FCM) using fluorochromes such as those belonging to the SYBR family (Chen et al. 2001; Marie et al. 1999; Middelboe and Glud 2003; Noble and Fuhrman 1998; Shopov et al. 2000; Wen et al. 2004). FCM can be used to perform very accurate and fast counts (Brussaard 2000; Li and Dickie 2001; Vives-Rego et al. 2000), generally in less than 2 min per sample (Marie et al.

1999). These last two points are important when a large number of samples have to be analyzed and statistically significant data are required. Unfortunately, as counting cannot always be done on the same day as sampling, reliable fixing and storage procedures may be a critical aspect.

A wide range of procedures, fixatives, dyes and storage has been proposed to date. Historically, considerable efforts have been made to develop procedures that accurately determine and preserve marine bacterioplankton and/or virioplankton in natural samples (Decamp and Rajendran 1998; Lebaron et al. 1998; Trousselier et al. 1995; Turley and Hughes 1992; 1994) or marine viruses in culture (Brussaard 2004; Marie et al. 1999); but only a few studies have been done involving freshwater samples (Lebaron et al. 1998). Some FCM studies have compared various preservatives and storage protocols (Marie et al. 1999; Troussellier et al. 1995; Turley and Hughes 1992), or the use of different nucleic acid stains (Chen et al. 2001; Lebaron et al. 1998; Tomaru and Nagasaki 2007). Different dilution solutions or incubation temperatures have been tested on virus counts by Brussaard (2004). Similar work has also been done for EFM (Ammini et al. 2010; Turley and Hugues 1992; Wen et al. 2004). Some studies have attempted to compare two of the three methods, typically either EFM *vs.* FCM (Gasol et al. 1999; Jochem 2001; Lemarchand et al. 2001) or EFM *vs.* TEM (Bettarel et al. 2000; Hara et al. 1991; Hennes and Suttle 1995; Noble 2001); but very few studies have attempted to compare the effectiveness of all three techniques (TEM, EFM, FCM) for performing direct total counts of bacteria and viruses (Chen et al. 2001; Marie et al. 1999). It would appear that the FCM counts were always correlated to, but slightly higher than those obtained by EFM or by TEM.

In this study, we chose to perform FCM tests with the most popular stains of the SYBR family, and various incubation temperatures, fixatives and dilution solutions and various storage conditions to optimize the counts of viral and non-photoautotrophic (commonly known as heterotrophic) bacterial communities sampled within the three largest natural French lakes (Annecy, Bourget, Geneva). Similar tests were done using EFM, and whenever possible, the two techniques were compared. Our study will highlight that FCM gives better results than EFM (suggesting that the concentrations of the total viruses in aquatic ecosystems are probably higher that many previous VLPs (Virus-like Particles) estimates found in the literature) but also that it may be problematic to believe that only one protocol can exist for all situations.

Materials and Methods

Sample Collection

Polypropylene bottles, previously rinsed with water from the collection sites were used to collect water samples, between September 2002 and November 2004 and during summer-fall 2012, from Lakes Annecy, Bourget and Geneva (details in Personnic et al. 2009). Immediately after sampling, water was filtered on board with <2 μm syringe filters to remove all potential grazers (and for keeping only free-living prokaryotes and viruses). Lake water so prepared and kept in cryotubes was then transported in less than 3 hours to the laboratory in refrigerated boxes (kept at 4-6°C to prevent any biological activity), and processed without delay for subsequent experiments. Data from Experiments 1-7 were used to test a wide range of dilution solutions, staining, fixation and storage conditions in FCM (see below). Experiment 8 was specifically designed to optimize EFM counts of bacteria and viruses (see below). Finally, the data obtained by EFM and FCM were compared. It is noteworthy here that we did not fix samples on board; differently said we did not test transport and storage conditions in affecting the abundances.

All nucleic-acid specific dyes were delivered in dimethyl sulfoxide and stored at -20°C. To avoid contamination, fixatives and dilution solutions, adjusted to a final pH of 7.5 were filtered through 0.2 µm (for bacterial analyses) or 0.02 µm (for viral analyses) using syringe filter (Millipore). Fixation was performed for at least 15 min at ambient temperature (*i.e.* ~20°C), and each sample was vortexed and then diluted 50 fold in the dilution solution. Incubation with the dye occurred at dim light at 20, 45, 65 or 75°C (see later).

Background

We only consider thereafter SYBR dyes that have proved to be very efficient and are the principal ones used nowadays for both freshwater and marine studies. We are aware however that some earlier studies demonstrated that other dyes, for instance Syto, Picogreen or Hoechst, performed very well in other freshwater ecosystems and that new dyes would also deserve to be tested such as the CF™ Dyes (from SIGMA). Originally, SYBR green dyes were preferred in marine studies as they are less sensitive to the salt, but they were also shown to work perfectly in freshwaters. Based on our experience (not shown), SYBR dyes can be used in a variety of freshwater ecosystems (from rivers to mountain lakes). It is possible however that in the case of some high-mountain lakes, none of the *Molecular Probes* green stains may be of any utility when there is glacial silt in the lakes (person. com.). More classical stains such as DAPI should thus be preferred. Also, our study is based on the use of a single laser beam fixed at 488 nm (blue) while less conventional instruments than the flow cytometer used in this study, with lasers at other and/or multiple wavelengths, can offer now the opportunity to use many other dyes (with maybe much higher resolution than the SYBR Green family dyes). Differently said, we are aware that this study is not complete and there is no pretention to give the only right solution. Our aim was more to end up with consideration (some possible tests) that should be taken into consideration to perform accurate counts of freshwater bacterio- and virioplankton communities by flow cytometry and epifluorescence microscopy.

It is also noteworthy here that we used the terms heterotrophic bacteria or bacterioplankton while heterotrophic prokaryotes should be preferred to take into account that FCM analyses cannot discriminate between Bacteria and Archea. As authors generally use “bacteria”, we also used preferentially this term, even if it is probably not *sensu stricto* true here.

Experiments Conducted

Experiment 1. This experiment was designed principally to choose among a wide panel of dilution solutions (TE, TAE, TBE, lake water, FACSFlow and PBS) for FCM analyses of heterotrophic bacteria. Samples were left untreated or fixed with formaldehyde (FA 1 or 2% final concentration, Sigma), glutaraldehyde (GA 0.25 or 1%, Sigma) or paraformaldehyde (PF 1 or 2%, freshly prepared home-made solution, (Marie et al. 1999), diluted in each dilution solution and finally stained with either SYBR Green I and II (Molecular Probes) for 15 min at room temperature to be analyzed by FCM.

Experiment 2. The type and the final concentrations of the fixative (GA 1 and 2%, PF 1 and 2%, FA 1 and 2% and PF 1% mixed with GA 0.05%) were tested, as were the storage temperature (4, -20 and -196°C) and duration (t = 0, 1, 8, 30 days) for bacterioplankton and/or virus FCM analyses. From experiment 1, we retained treatments that gave the best statistical results for both cell abundance and FCM signatures (*i.e.* TAE, TE, TBE and lake water among the dilution solutions and SYBR Green I for staining). Briefly, 10 mL subsamples were immediately treated with one concentration of a given fixative. Another subsample was kept without fixative. From each of these subsamples, a set of 1 mL duplicate subsample was kept at 4°C and was then analyzed on days 0, 1, 8 and 30, and 3 sets of 1 mL duplicate subsamples were left for 15 min at 4°C before being stored at –20°C or dove in liquid nitrogen. In this way, each frozen sample was thawed only once just before being analyzed at t = 1, 8 or 30 days. Previous results suggested that repeated thawing clearly has a negative impact on the total counts, and we obviously wanted to avoid this (data not shown). At t0, the samples were diluted as seen above. At t1, 8 and 30 days, refrigerated samples, particularly those stored at –20°C and liquid nitrogen, were warmed to room temperature over 15 min (previous experiments had shown that cold samples had a negative effect on the quality of the FCM signature), vortexed and diluted. The diluted samples were then stained for 15 min at room temperature with the nucleic acid dye SYBR Green I at a final concentration of 10^{-4} for the bacteria and $5x10^{-5}$ for the viruses (Marie et al. 1997).

Experiment 3. The staining characteristics and in particular the staining kinetics of SYBR Green I and SYBR Gold, both at a final concentration of 10^{-4}, were analyzed using bacterioplankton samples which had been fixed with GA 2%. Samples were diluted in TE buffer, and replicate samples were analyzed by FCM at regular intervals after incubation with the dye for 1 min to 1 h.

Experiment 4. This experiment was performed to compare the staining efficiencies of SYBR Green I (final concentration of 10^{-4}) and SYBR Gold (final concentration of 10^{-4}, 5 x 10^{-5}, 2 x 10^{-5}) for counting bacteria and viruses by FCM. Briefly, 12 samples which were fixed to a final concentration of 1% GA were diluted in TE or lake water and analyzed by FCM after incubating with the stain for 15 min at 20°C for bacteria, or after incubating at 45°C, 65°C or 75°C for viruses according to Marie et al. (1999) and Brussaard (2004).

Experiment 5. This experiment aimed to test the use of fixatives and dilution solutions on FCM virus counts. Samples were fixed with FA, or GA (1% final concentration) or left untreated and dilution was performed in lake water, in TE and in TBE. All samples were stained with either SYBR Green I or SYBR Gold, at a final concentration of 10^{-4}. Incubation was performed at 75°C (Marie et al. 1999).

Experiment 6. A range of incubation temperatures was also tested. Samples were either left unfixed or fixed with 1% or 2% GA and analyzed at day = 0 or at day = 1 after storage at 4°C or -20°C. For FCM analyses, samples were diluted in TE or in lake water and stained with SYBR Green I (final concentration of 10^{-4}) and subjected to various incubation temperatures 20°C (15 min), 45°C and 75°C, the latter two lasting for 10 min, followed by cooling for 5 min at room temperature.

Experiment 7. In order to test different ways of using TE for FCM virus counts. The conditions tested included the final pH of the TE solution (pH 7 or 8, corresponding to the natural pH variation observed in our lakes), TE was autoclaved or not, filtered through a 0.02-µm or 0.2-µm filter. The samples were then diluted in the various types of TE buffer and incubated with SYBR Green I or SYBR Gold, both used at a final concentration of 10^{-4}, for 10 min at 75°C, and FCM analyses were done. It is noteworthy here that each test was also performed on a "negative" control (*i.e.* without sample).

Experiment 8. Tests for EFM counts were done by counting both bacterio- and virioplankton in several water samples. Samples were either left untreated or immediately fixed upon arrival in the laboratory for at least 15 min. FA or GA, were added at final concentrations of 1 and 2% or 0.5 and 1%, respectively. 1 mL of the fixed samples was filtered through a 25-mm, 0.02-µm ultra-fine pore size filter (Anodisc, Fisher Scientific), backed by a 25-mm GF/C filter (Whatman), at low vacuum. Each filter was stained either with SYBR Green I (Noble and Fuhrman 1998) or with SYBR Gold (Chen et al. 2001), both at a final concentration of 10^{-3}. The filter was finally mounted on a glass slide and 30 µl of an anti-fading solution was added, that was prepared as recommended by Noble (2001). Briefly, we prepared a solution of 990 µl of a 50% PBS 50% glycerol mixture with 10 µL of 6.6% p-phenylenediamine. Then, a cover slip was placed over the filter. As suggested by Wen et al. (2004), the slides were prepared immediately, and if possible counted on the same day or else stored at – 20°C for no more than a couple of days. In addition to the viral and bacterial counts at t = 0, 10 of the filters (5 filters for bacterial counts, 5 for viral counts), were re-counted after being stored for 16, 29, 53, 68, or 96 days at –20°C in order to assess the importance of counting the slides immediately after they had been prepared. The results have been expressed as a percentage of the counts obtained at t = 0.

Comparison of FCM and EFM

Samples were collected monthly from February to May 2003 from several depths (between 2 and 50 m) in the three lakes. A total of 80 samples were analyzed for bacteria and viruses in order to compare the counts obtained by FCM and EFM. For the FCM analyses, samples were fixed in 0.5% GA, and stained with SYBR Green I at a final concentration of 10^{-4}. Samples for bacterial counts were incubated at room temperature, and samples for viral counts were heated for 10 min to 75°C and then cooled for 5 min prior to analysis. For EFM analyzes, the samples were fixed with 1% FA and incubated with SYBR Gold at a final concentration of 10^{-3}. The slides were prepared without delay and either counted immediately or stored at -20°C for up to a few days.

Flow Cytometry Analyses

The counts were done using a FACSCalibur (Becton Dickinson) benchtop flow cytometer, equipped with a blue laser beam fixed at 488 nm and with the standard filter setup. The main FCM procedures were the same as those outlined by Marie et al. (1999), and originally devised for marine bacterioplankton and virioplankton. MilliQ water was used as a sheath fluid since no significant differences were recorded with either filtered lake water or the FACSFlow provided by the manufacturer (not shown). Samples were run at medium speed (*i.e.* between 60 and 70 $\mu l.min^{-1}$, the flow rate being checked before each analysis), and all parameters were acquired in log mode. To avoid coincidence, the number of events was limited to between 100 and 1000 per sec by diluting the sample and/or by raising the threshold of the instrument. Bacterial cell parameters were determined relative to the values found for an internal standard, *i.e.* a solution of 1-µm fluorescent beads (Molecular Probes). Bacteria and viruses were detected from dot plots of right angle light scatter (SSC) versus the green fluorescence of the acid nucleic dye complex (FL1 channel: 530±15 nm) and the red fluorescence of phytoplankton (FL3 channel: >630 nm) versus FL1. Data were collected in listmode files and then analyzed on a separate PC using the custom-designed freeware CYTOWIN (Vaulot 1989). Abundances are reported as $cells.mL^{-1}$ (heterotrophic bacteria) or $particles.mL^{-1}$ (viruses).

Epifluorescence Counts

Counts were done using a LEICA epifluorescence microscope equipped with a mercury lamp and a blue excitation light (450-490 nm). Around 200 bacterial cells (cyanobacteria were excluded from this counting) were counted in 10 randomly selected fields for each filter, and 400-600 viruses were counted in 20 fields. The viral and bacterial abundances have been reported as $particles.mL^{-1}$ or $cells.mL^{-1}$ respectively, following the procedures outlined by Noble (2001).

Statistics

Bacterial and viral concentrations we obtained following the different treatments were compared and analyzed for significance by using the tests of Mann-Whitney or Kruskal-Wallis with the PAST software package (freely available at http://folk.uio.no/ohammer/past/).

Results

Choice of the Nucleic Acid Stain for FCM Counts

When the dyes were used at a final concentration of 10^{-4}, compared to SYBR Green I stained samples, SYBR Green II yielded 40% less bacterial numbers (Figure 1, experiment 1). This figure also shows that the SYBR Green II counts were correlated to the SYBR Green I counts. The fluorescence of the SYBR Green I stained samples reached maximum and stable bacterial abundances after having been incubated with the dye for 10-12 min, whereas the fluorescence of those stained with SYBR Gold increased less rapidly and was less stable (experiment 3). In the latter experiment the SYBR Green I stained samples showed 16% greater abundances than the SYBR Gold stained ones. From a qualitative point of view, the bacterial signature was easier to interpret when SYBR Green I stain was used (not shown). The 4th experiment, in which we counted bacteria and viruses within various water samples, gave us an indication of the staining efficiency of SYBR Green I compared to that of various concentrations of SYBR Gold.

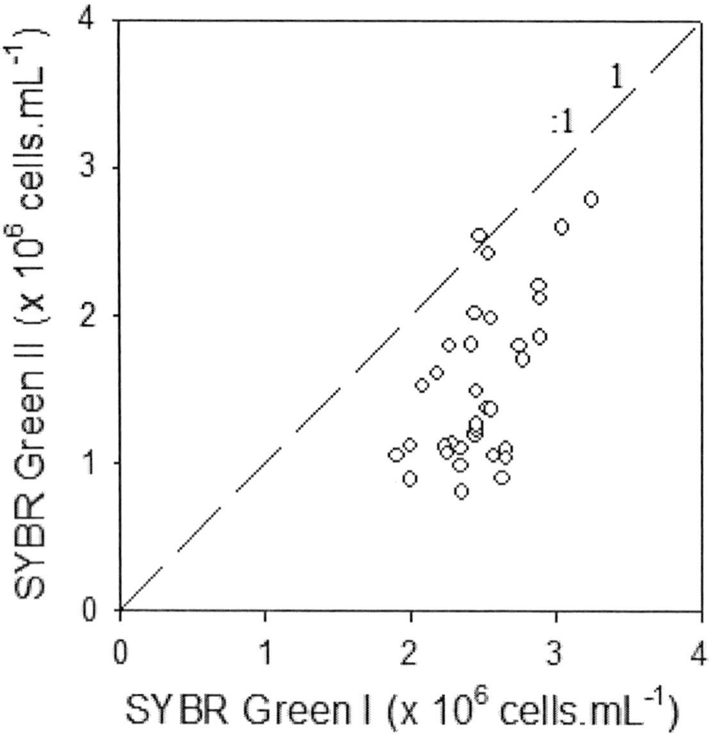

Figure 1. Relationship between SYBR Green I and SYBR Green II, both used at a final concentration of 10^{-4}, for bacterioplankton-stained samples diluted using various dilution solutions (TAE, TE, TBE, lake water, FACSFlow, PBS), fixed with different types and concentrations of fixatives (FA, GA, PF) and analyzed by FCM. $y = 1.11x - 1.23$ ($n = 34$, $r = 0.6$, $p = 0.99$). The dashed line corresponds to the 1:1 relationship. Experiment 1.

Average total bacterial cell count was not significantly different for SYBR Green I used at a final concentration of 10^{-4} and SYBR Gold used at one of the range of concentrations (10^{-4}, 5×10^{-5} or 2×10^{-5}, data not shown).

For virus counts staining with SYBR Green I used at a final concentration of 10^{-4}, rather than with SYBR Gold used at concentrations of 10^{-4}, 5×10^{-5} or 2×10^{-5}, gave virus concentrations that were significantly lower (-28%, Figure 2).

The high standard deviations led us to conclude that the samples were very heterogeneous. In addition, the temperature of incubation had a critical role on viral staining efficiency when SYBR Green I stained (10^{-4}) with mean virus abundances being significantly lower at 45 than at 65 or at 75°C. From a qualitative point of view, the use of SYBR Gold was preferable to SYBR Green I and the lower the concentrations of SYBR Gold, the greater the number of detectable subpopulations within the viral community, whatever the temperatures of incubation (Figure 3).

At 75°C, up to 5 populations could be detected using the lowest concentration of SYBR Gold, versus only 3 populations detected using the highest concentration of SYBR Gold or SYBR Green I.

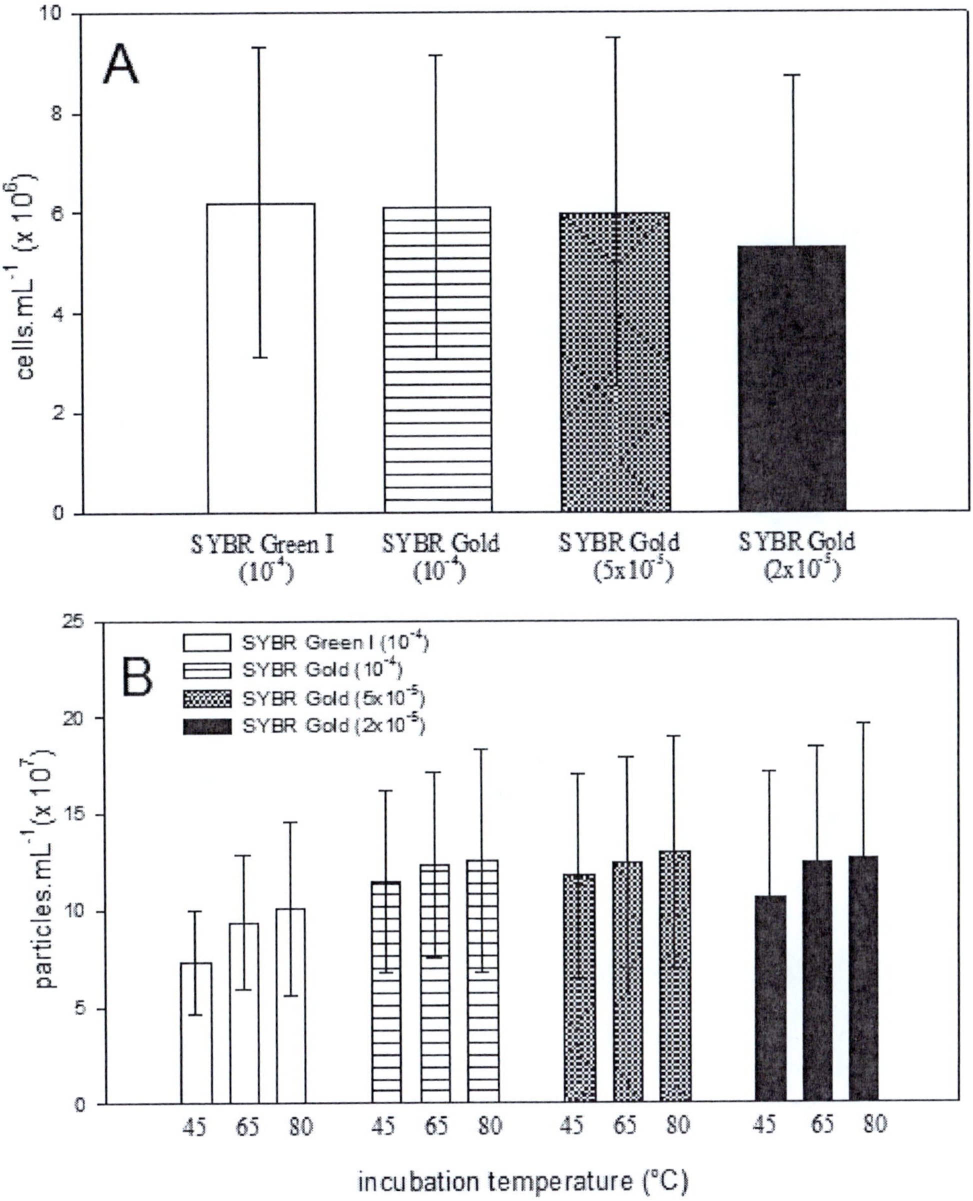

Figure 2. FCM counts of heterotrophic bacteria (A) or viruses (B) stained with SYBR Green I (10^{-4}, white bars) or SYBR Gold at 3 different concentrations (10^{-4}, hatched bars; 5 x 10^{-5}, dotted bars; 2 x 10^{-5}, black bars) for 12 different samples. Viruses were incubated with each dye and concentration tested at 3 different incubation temperatures (45°C, 65°C and 75°C). Error bars are relative to 12 different water samples. Experiment 4.

The results of the 5th experiment, showed again that SYBR Gold and SYBR Green I counts in different water samples, correlate very closely and positively ($n = 98$, $r = 0.42$, $p = 0.99$ for viruses, $n = 65$, $r = 0.94$, $p = 0.99$ for bacteria). In Experiment 4, virus counts were still correlated, but were significantly lower when SYBR Green I was used rather than SYBR Gold (about 20% lower in the case of the 75°C series). Such differences were clearly confirmed here.

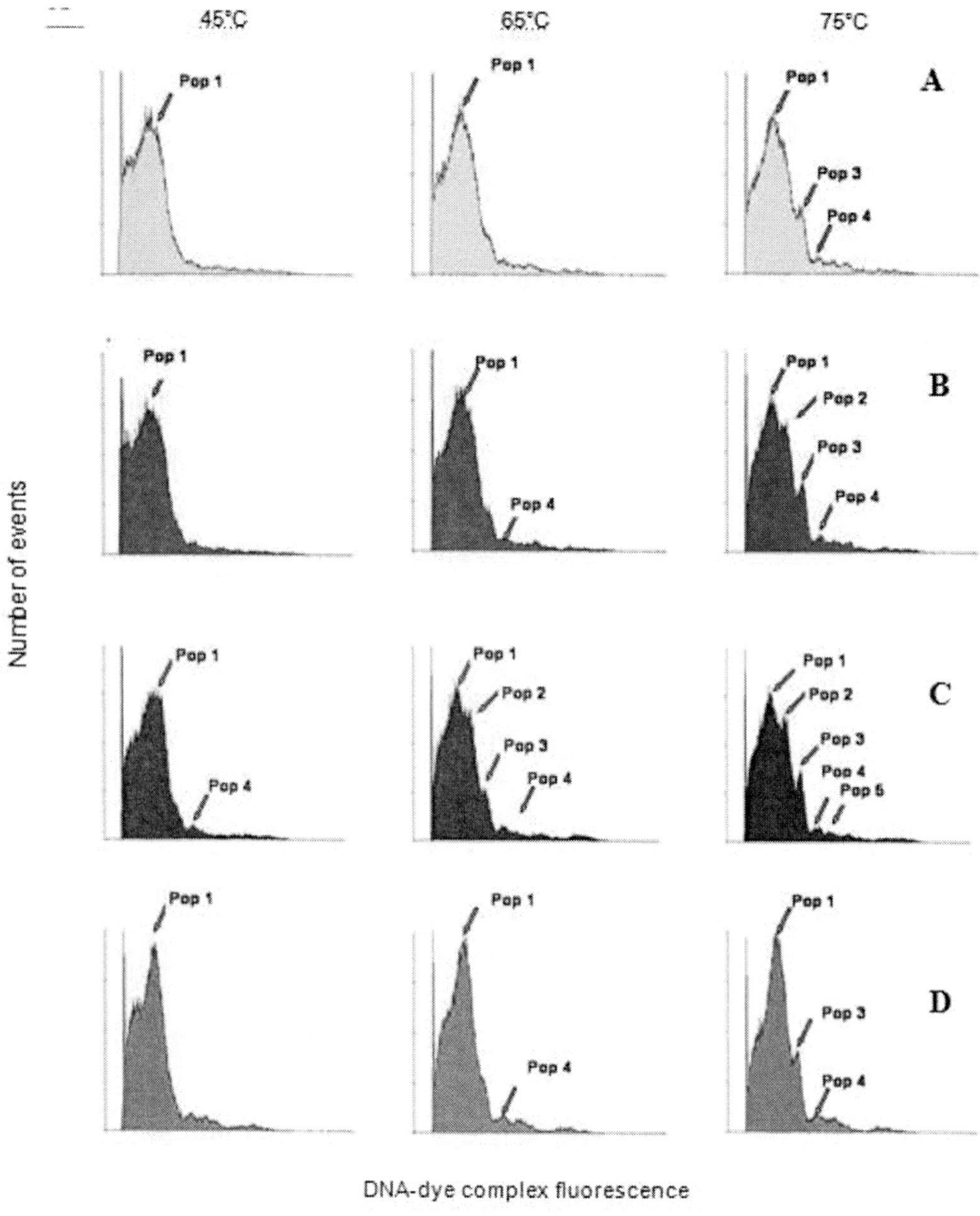

Figure 3. Histograms of virus distributions showing different populations or groups (Pop). Samples were stained with SYBR Gold at a final concentration of 10^{-4} (A), 5 x 10^{-5} (B) or 2 x 10^{-5} (C) or with SYBR Green I at a final concentration of 10^{-4} (D), and incubated at different temperatures (45°C, 65°C or 75°C) for 10 minutes. Experiment 4.

The Choice of the Dilution Solution for FCM Counts

Throughout the first experiment, FACSFlow and PBS used as dilution solution provided the lowest bacterial concentrations, 16% (SYBR Green I) and 52% (SYBR Green II) less than when samples were diluted in TE. From a qualitative point of view, TE, TAE and TBE allowed to distinguish between different populations (Figure 4A).

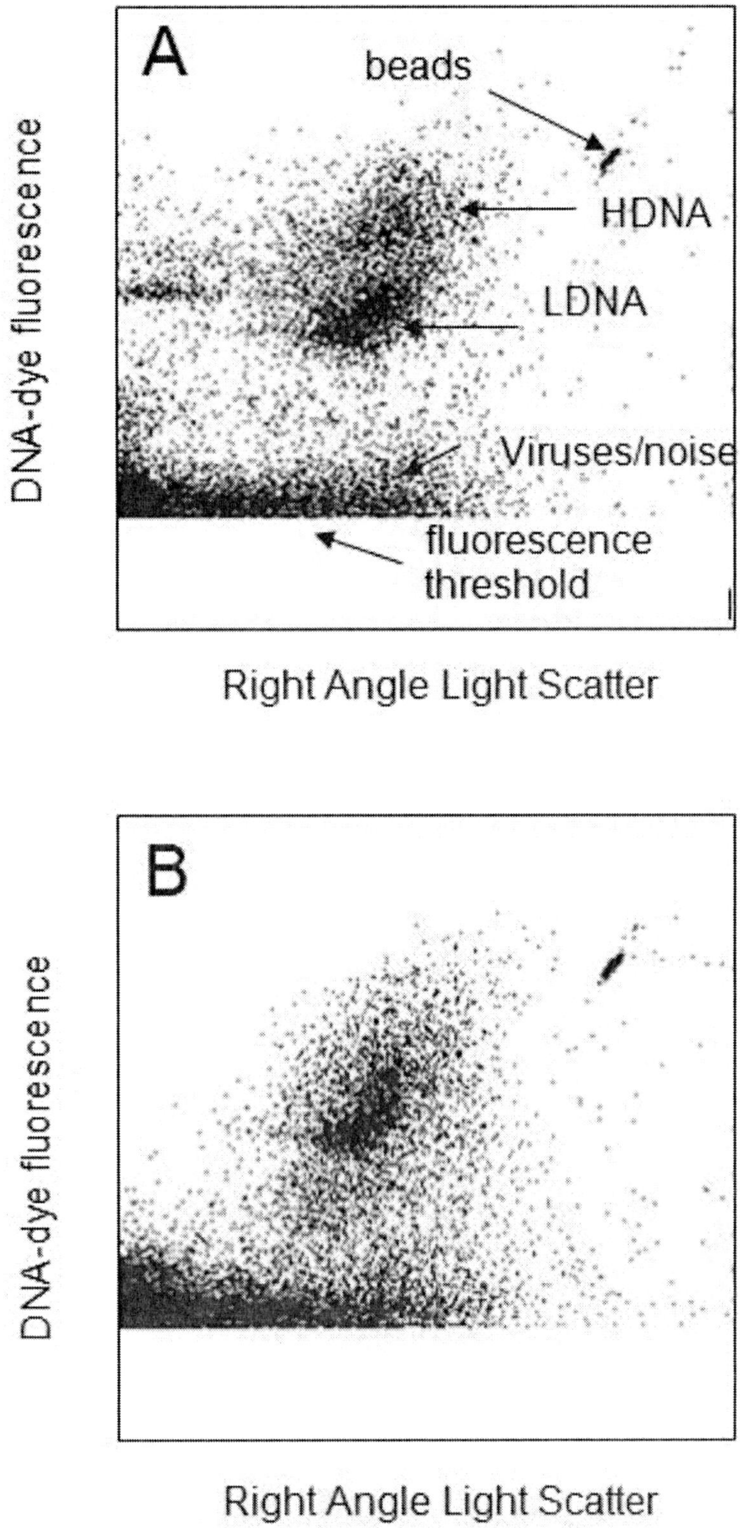

Figure 4. Typical cytograms obtained for bacterioplankton analysis using 0.2 μm filtered TE (A) or lake water (B) to dilute samples.

By comparison, filtered lake water furnished the most compact signatures in combination with unfixed samples (Figure 4B). In addition to the findings of the second experiment, no significant quantitative differences were observed between TE, TAE, TBE or lake water at t0 (data not shown). At t1, t8 and t30, the results were surprisingly different. Bacterial counts were significantly higher when samples that had been stored at 4°C or at –20°C, were diluted in TE or in lake water, than when they were diluted in TAE or TBE. For instance, TBE dilution gave values up to 15% lower than when TE was used. Considering all the viral counts, regardless of whether SYBR Green I or SYBR Gold stain was used, no significant differences were observed after diluting in TE or in lake water. Nevertheless, as for the bacterial counts, TE made it possible to distinguish between various viral subpopulations. Using autoclaved or non-autoclaved TE did not have any influence on the total bacterial counts found (experiment 3, data not shown). This was not the case for viruses (experiment 7). Indeed, when using non-autoclaved TE buffer, some virus populations, which are situated at lowest fluorescence values within the flow cytogram, could overlap with the background noise (corresponding to debris and electronic noise) (Figure 5A, B). Using autoclaved TE buffer circumvented this problem by reducing the noise and by somehow shifting the background noise away from the virus population signatures (Figure 5C). Not to autoclave the TE buffer did significantly influence the total virus counts obtained. The overlapping of the distributions of both signal and cytometric noise fluorescence resulted in possible overestimations of ca. 30% of the viral population 1 (VLP1) and referred to as the bacteriophage community, situated at the lowest fluorescence values within the cytogram (Personnic et al. 2009). When analyzing the controls, we found that the noise within autoclaved controls was reduced up to 7 fold compared to that within the non-autoclaved controls. Total virus counts were not affected by using TE at a pH 7 or 8, nor when filtering the recently made buffer through either 0.2 or 0.02 µm.

Choice of the Fixative for FCM Counts

Experiments 1 and 2 provided useful results concerning the choice of the fixative for bacterial counts (not shown). At t0, the smallest numbers of bacteria were recorded for non-fixed samples, whereas the use of fixatives increased significantly the number of detectable bacteria by an average of 14%. The highest number of bacteria at this time was obtained with a 2% final concentration of GA for all buffers. Up to 34% (on average 21%) more bacteria were detected in GA 2% fixed samples than in fresh ones. In our experiments, GA 1 or 2% gave the highest counts. From a qualitative point of view, fixing sometimes made it possible to distinguish between different bacterial populations, even in filtered lake water. For virus counts, we did not observe any quantitative or qualitative difference weather fixing the samples or leaving the sample unfixed (experiment 5, not shown).

Storage Conditions for FCM Counts

For unfixed samples, a significant increase occurred in bacterial abundance after storage for 8 days at 4°C, with concentrations that could be up to 8 times higher than at t0 (Figure 6A).

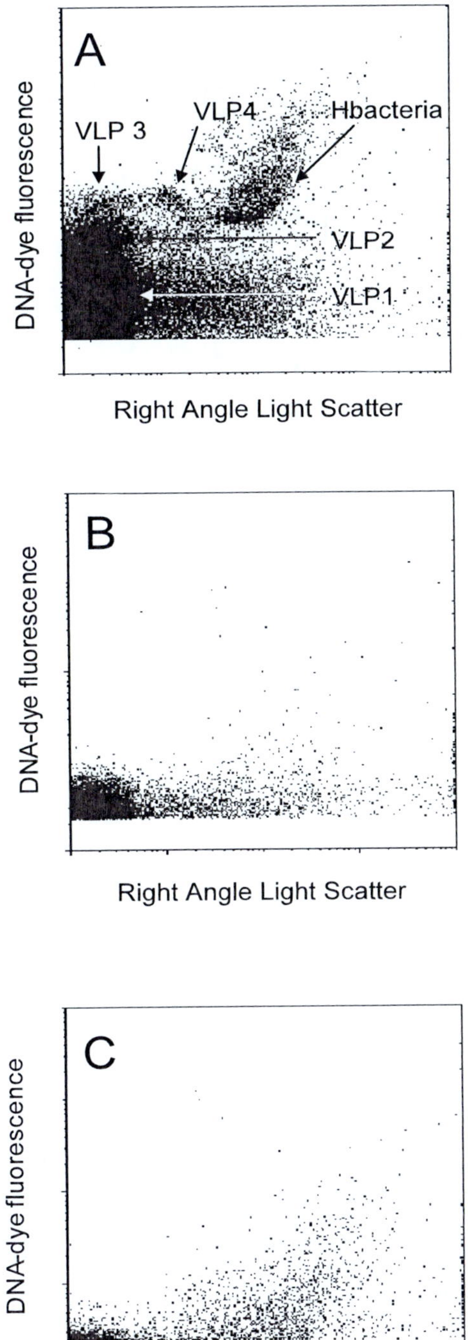

Figure 5. A: Typical FCM cytogram, representing both the heterotrophic bacterial (Hbacteria) community and different viral populations (VLP) stained with SYBR Gold (10^{-4}). B: Control cytogram with no sample and showing the signature of non-autoclaved 0.02 µm filtered TE stained with SYBR Gold (10^{-4}). C: Control cytogram with no sample, showing the signature of autoclaved TE stained with SYBR Gold (10^{-4}). Experiment 7.

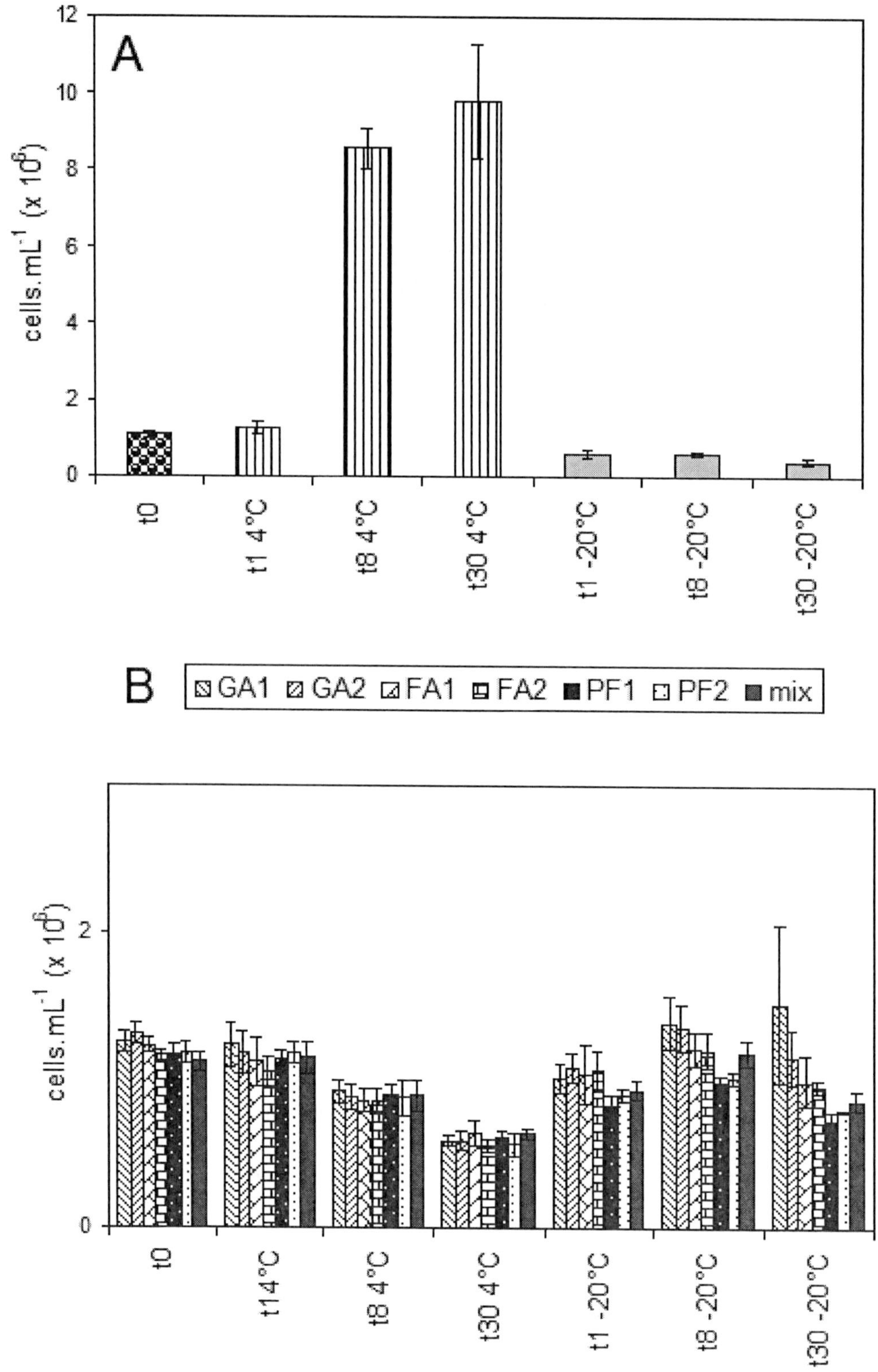

Figure 6. (Continued).

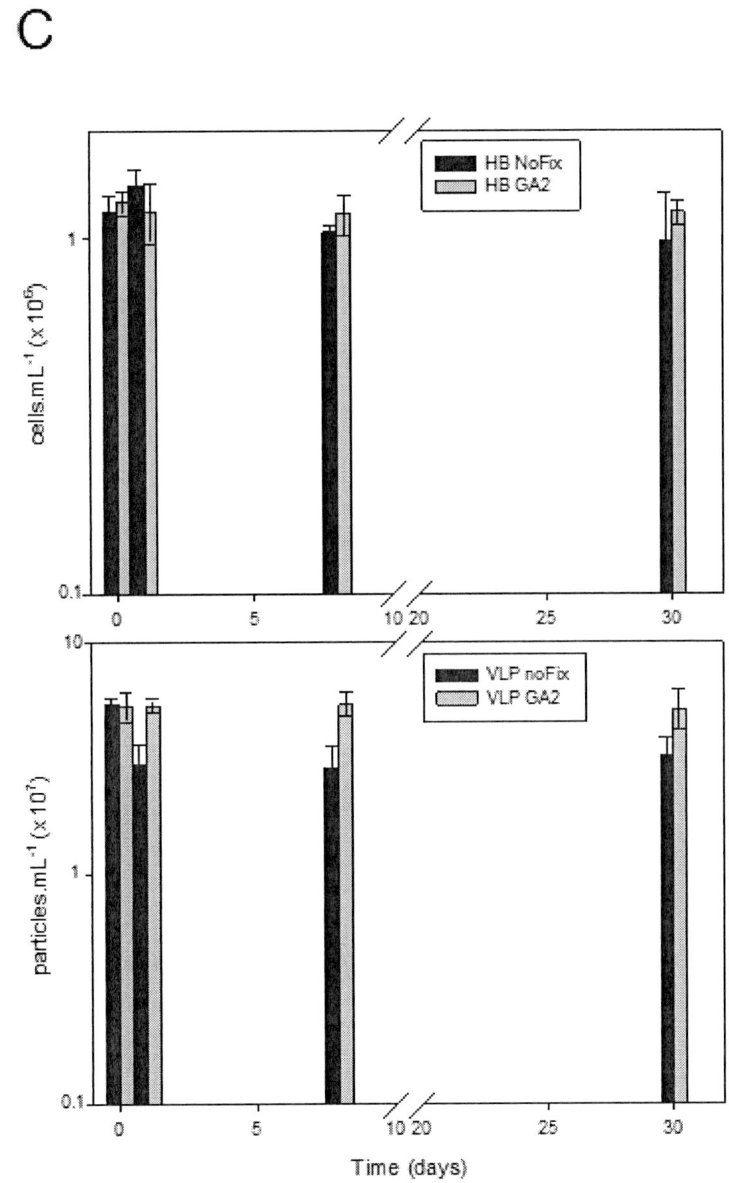

Figure 6. FCM for bacterioplankton and/or viral samples. A: Unfixed samples were analyzed at t = 0 (dotted bar) and at t1, t8 and t30 after being stored at 4°C (hatched bars) or –20°C (gray bars). B: Samples fixed with different fixatives (GA, FA, PF, mix: PF1 and GA 0.05%) and at different concentrations (1 or 2 %) were analyzed at t = 0, and at t1, t8 and at t30 after being stored at 4°C or –20°C. C: Unfixed and fixed samples with 2% glutaraldehye, flash-frozen in liquid nitrogen and analysed after 1, 8 or 30 days. Experiment 2.

At t30, bacterial concentrations were up to 10 times higher than at t0. The small increase at t1 compared to t0 was not significant. At –20°C, abundances decreased in a significant way, *i.e.* at t1 (and at t8), and then at t30 these non-fixed, frozen samples showed a decrease in the initial abundance by about 46% and 66%, respectively.

Figure 6B, referring to the fixed samples, clearly shows that at 4°C a gradual and significant decrease in the initial total bacterial counts occurred from t0 (or t1) to t8, and from t8 to t30, by 23% and 50%, respectively. When stored at -20°C, the concentrations found were significantly lower than those found at t0 for the sets which were thawed both at t1 and at t30 (-20%).

The second frozen set analyzed at t8 did not display any significant change in counts compared to t0 (Figure 6B). When samples were flash frozen (in liquid nitrogen) without fixation, both bacteria and viruses decreased markedly while with fixation (either 1 or 2% glutaraldehyde), no significant differences were recorded for both communities between t0 and t1, t8 or t30 (Figure 6C).

Dye Incubation Temperature for FCM Counts

The results of the incubation temperature experiment for bacterial analyses (experiment 6) have been illustrated in Figure 7. In unfixed samples the number of bacteria detected by FCM decreased significantly at temperatures above 45°C, at 75°C the number of total bacteria being reduced by an average of 22% compared to data obtained at 20°C or at 45°C. For heated and unfixed samples, cell losses were on average greater when samples were diluted in lake water.

Fixing the samples with either GA 1 or 2% yielded significant higher counts than unfixed samples especially at higher temperatures. The efficiency of detecting GA 1 or GA 2% fixed cells was not significantly different if they were heated to 45°C or 75°C, except for the samples diluted in lake water and heated to 75°C, for which we found significant lower concentrations. At temperature exceeding 45°C it appeared that samples which were diluted in TE rather than in lake water were more "protected" from overheating and cell destruction. The results of the incubation temperature experiment for virus samples (experiment 4) have been illustrated in Figure 3. Temperature was proved to be of great importance in the discrimination and the counting of viruses. Two, 3 and 5 viral groups were detected at 45°C, 65°C and 75°C, respectively. At 65 or 75°C, virus counts were significantly higher (+14%) than those at 45°C.

EFM Counts

Each filter was analyzed both for bacteria and viruses. SYBR Gold and SYBR Green I counts ($n = 28$, $r = 0.87$, $p = 0.99$) were positively correlated and showed no significant quantitative differences. The bacterial concentrations found with SYBR Green I and SYBR Gold ranged between $4.28 \times 10^5 - 1.76 \times 10^6$ and $1.84 \times 10^5 - 2.56 \times 10^6$ cells mL^{-1}, respectively.

Viral concentrations displayed a range of $1.09 \times 10^7 – 5.43 \times 10^7$ viruses.mL^{-1} with SYBR Green I, and of $5.01 \times 10^6 – 5.73 \times 10^7$ particles.mL^{-1} with SYBR Gold. At 10^{-3}, SYBR Gold yielded a more stable fluorescence than SYBR Green I. No obvious trend could be discerned related to whether different fixative solutions had been added. As illustrated in Figure 8, the time for which filters can be kept and still yield reliable bacterial and viral counts seems to be limited to 1 month. After 16 days, the estimates were similar to those at time zero (immediately after slide preparation).

After one month storage, there was an important decrease in abundance, estimated to be of 5% to 98% for viruses and 3% to 73% for bacteria. The decrease in viruses occurred faster than that in bacteria.

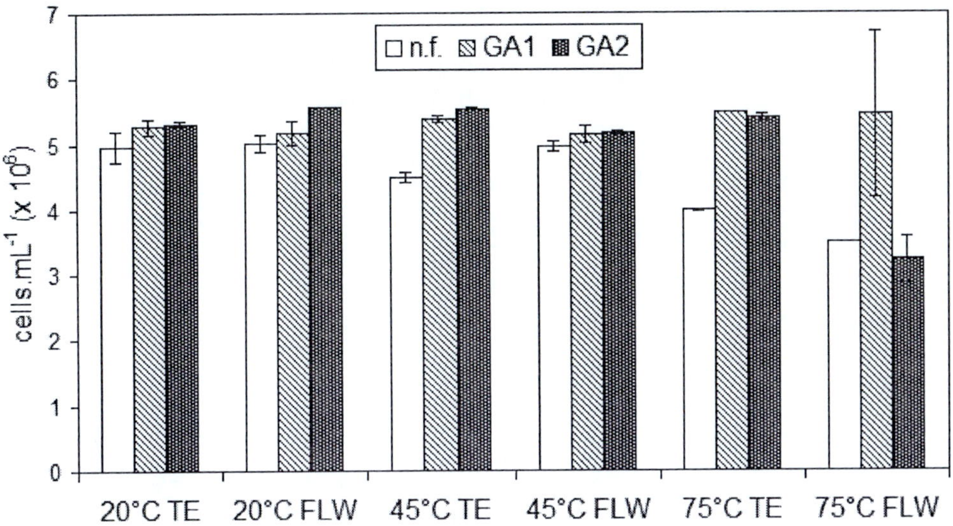

Figure 7. FCM bacterial counts at t = 0. Very similar results were obtained at t=1 (not shown here). Error bars represent standard deviations of duplicate counts. The samples were fixed in GA 1 or 2% or not fixed (n.f.), were diluted in 0.02 µm filtered TE or lake water (FLW) and incubated at temperatures of 20°C, 45°C or 75°C. Experiment 6.

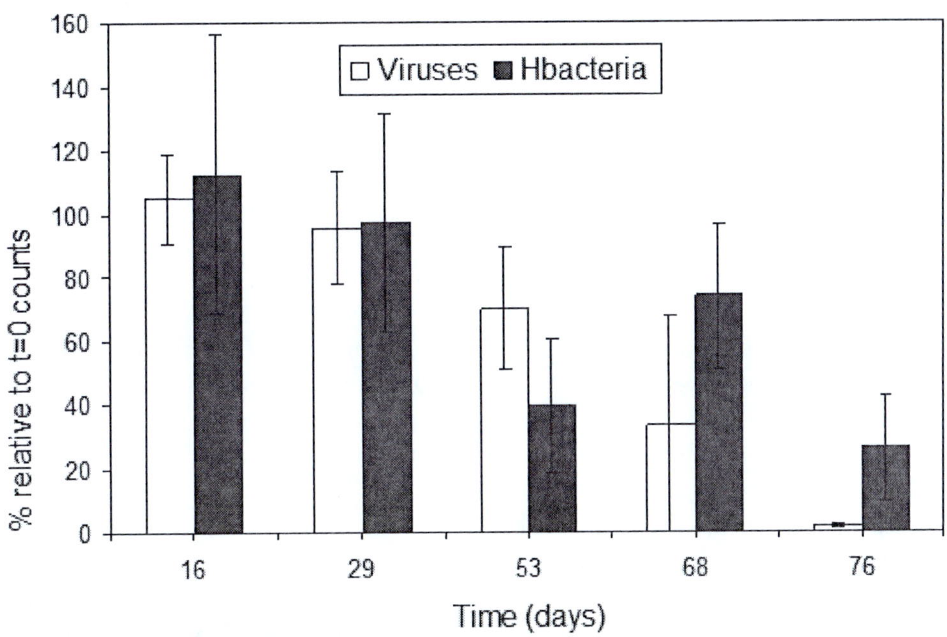

Figure 8. Percentage of bacterial and viral abundances determined by EFM after keeping the sample at –20°C for 16, 29, 53, 68 and 96 days, compared to the values obtained at t = 0. Error bars represent standard deviations of different samples (n = 5). Experiment 8.

Comparison between FCM and EFM Counts

The FCM counts were closely correlated to the EFM counts for both bacteria and viruses (Figure 9A and B). However, with FCM, bacterial estimates were 54% higher and viruses estimates were 32% higher than with EFM. Bacterial counts obtained using EFM ranged from 4.54 x 10^5 to 2.88 x 10^6 (mean 1.19 x 10^6 cells.mL^{-1}), and from 9.48 x 10^5 to 9.36 x 10^6 (mean 2.53 x 10^6 cells.mL^{-1}), using FCM. Viral abundances ranged from 5.54x10^6 to 5.71x10^7 (mean 3.36 x 10^7 particles.mL^{-1}) by EFM, and from 2.7 x 10^7 to 1.32 x 10^8 (mean 4.96 x 10^7 particles.mL^{-1}) by FCM.

Discussion

FCM Analyses

Our results indicate that bacterial and viral counts are quantitatively and/or qualitatively affected by the type and the final concentration of the fluorescent nucleic acid dye used, the incubation temperature and time, whether fixatives and dilution solutions are used and by the storage condition.

Total bacteria counts were highest with SYBR Green I (10^{-4} final concentration) and lowest (40% less) with SYBR Green II (same concentration). These results are not very surprising, as SYBR Green II, unlike the other two dyes tested, preferentially stains single-stranded DNA or RNA, rather than double-stranded DNA, which is the main form present in bacterial cells (indications by the manufacturer). As shown by Lebaron and co-authors (1998) and by our results, bacterial counts obtained by staining the samples with SYBR Green I were closely correlated to those obtained after staining with SYBR Green II. When the staining efficiency of SYBR Green I (10^{-4}) was compared to that of SYBR Gold (10^{-4}, 5 x 10^{-5} or 2 x 10^{-5}) on different natural samples (experiment 4), the mean bacterial abundances found did not differ significantly. On the contrary, when SYBR Gold and SYBR Green I (10^{-4}, in both cases) were tested on replicate water sample (experiment 3), SYBR Green I gave on average 16% higher abundances than SYBR Gold.

These two apparently contradictorily findings indicated that different results may be obtained for different water samples. In addition to the quantitative advantage of using SYBR Green I for bacterial counts, this stain also provides bacterial signatures, which were easier to interpret, especially in organic material rich water samples (data not shown). The kinetics experiment showed that SYBR Green I reached maximum bacterial abundances after only a few minutes, and 15 min was a good compromise before FCM analysis. With regard to the dilution solutions, we strongly advise against using FACSFlow or PBS. They both yielded significantly lower bacterial counts than TE, TAE, TBE or filtered lake water (-16%). Interestingly, at t = 0, no quantitative difference was found between the last 4 dilution solutions mentioned above. At t1, t8 and t30, TE and filtered lake water provided 15% higher abundances than TBE or TAE. From a qualitative point of view, TE, TAE and TBE allowed us to distinguish some bacterial subpopulations, typically two groups that had clearly differing DNA-dye fluorescence. These two groups had already been reported by Gasol et al. (1999) and by Li and Dickie (2001).

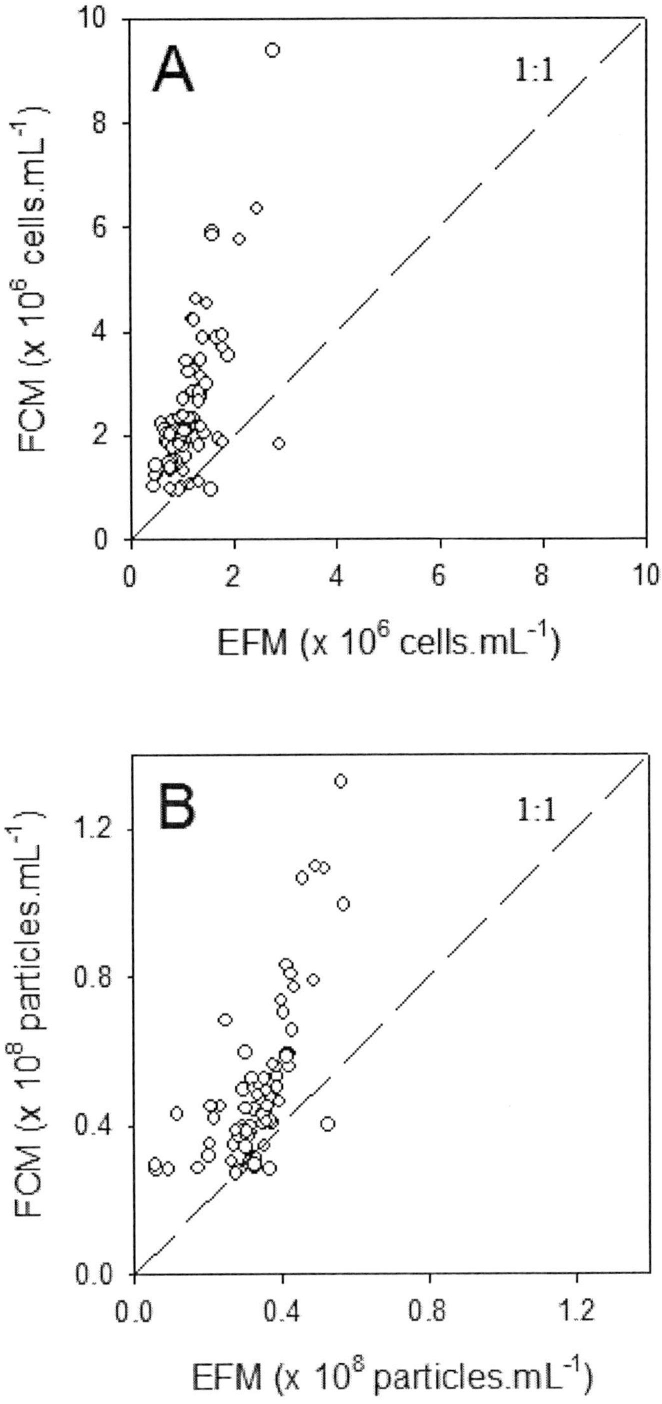

Figure 9. Relationships between bacteria (A) and viral (B) counts assessed by EFM and by FCM. The dashed lines correspond to the 1:1 relationship. A: $y = 1.50x - 0.0074$ ($n = 80$, $r = 0.7$, $p = 0.99$). B: $y = 2.08x + 0.07$ ($n = 80$, $r = 0.69$, $p = 0.99$). See Methods for the experimental conditions used.

They named them HDNA (for high DNA containing cells) and LDNA (for low DNA containing cells), or type I and II, respectively. Bacteria belonging to the HDNA or to type-I group are thought to be metabolically more active than those in the LDNA or type-II group (Gasol et al. 1999; Lebaron et al. 2002) although more recent studies highlighted that such a discrimination was probably not so clear (Bouvier et al. 2007). When samples were diluted in filtered lake water, the signal was generally more compact than when they were diluted in TE, TAE or TBE, likely to be due to the presence of EDTA in the Tris-buffers, which may interact with nucleic acid chains. Sometimes the use of fixatives had a similar effect on the signal, making it possible to distinguish between major subpopulations. One possible explanation for this may be that fixation can sometimes change the refractive index of the cell by affecting the right angle scatter, as well as DNA characteristics and thus fluorescence. At t0, regardless of the fixative used, bacterial abundances were 14% higher for the fixed samples than for fresh samples. This has also been reported by Marie et al. (1999). We may not exclude whether DNA populations observed were artifacts when a given buffer or fixative was used or not.

Generally speaking, fixatives are used to avoid the occurrence of significant changes in the cell counts and characteristics over time. Moreover, fixatives (and also heating treatments), may make the cells more permeable, allowing high-molecular weight molecules (such as the specific nucleic acid stains) to penetrate the cells more quickly and easily (Lebaron et al. 1998, Marie et al. 1999). We tested some members of the aldehyde family (FA, GA, PF), as they are known to penetrate cells rapidly, because of their relative low molecular weights (Hayat 1970; Xenopoulos and Bird 1997). FA is known to crosslink proteins within the cell membrane, and to influence cell morphology (Noble 2001; Vaulot et al. 1989). PF is the polymerized form of FA and unlike FA, PF lacks cross-linking characteristics (Marie et al. 1999). If fixation affects the cell morphology, the forward angle scatter which is related to the size of the cells may also change, thus modifying the signal recorded by FCM (Navaluna et al. 1989). GA is usually used in electron microscopy studies, as the cell shape is little changed even if the stain produces cross links with cell proteins (Vaulot et al. 1989). In our study, GA used at a final concentration of 1 or 2% seemed to be the most appropriate type of fixative. When unfixed or fixed samples were stored at 4°C, abundances found at t = 1 were found to be similar at t = 0, suggesting that analysis could be postponed by one day (see also Jacquet et al. 1998). At 4°C and in unfixed samples, counts dramatically increased between t1 and t8, indicating a rapidly-growing community despite the low temperature. At -20°C, these unfixed samples showed an undoubted decrease in counts, -46% and –66%, after 1 day and 1 month of storage, respectively. One hypothesis is that at very low temperatures and without a gradual temperature decrease, unfixed cells encounter physical problems (*e.g.* intracellular freezing) that result in cell damage. These considerations obviously lead us to discourage the storage of unfixed samples (that was also confirmed when using liquid nitrogen without previous fixation). Then, what occurred when the samples have been fixed? At 4°C, we detected a loss of total abundance at t8 and t30 by 23 and 50% respectively; no loss was detected at t1. When fixed samples were stored at -20°C, we noticed that the concentrations for the sets which have been thawed at t1 and at t30 were significantly lower (by 20%) than the values t0. Generally speaking, a loss in cell numbers may be due to several factors, such as attachment to the wall of the recipient or burst due to virus infection (Turley and Hughes 1992). Cells may encounter uninhibited enzyme activity (Gundersen et al. 1996) causing cell dissolution, or cells may break due to inappropriate physical (temperature) or chemical (fixation) conditions. Gundersen et al. (1996) suggested that major bacterial losses may occur as a result of uninhibited protease activity, even in fixed water samples. They found bacterial losses of 5% and 50% after 9 and 29 days of storage respectively at -20°C for samples fixed with 2.5% GA. Brussaard (2004) demonstrated that a one month storage at 4°C or -20°C of samples fixed in 0.5% GA led to considerable reductions of viral abundance. Her findings must also be applicable to the storage of bacterioplankton samples. Turley and Hughes (1992) also reported a significant decline in bacterial counts when they analyzed bacterioplankton samples fixed in 1% GA and which had been stored at room temperature – cell numbers were down to 39% of the initial counts prior to storage. Trousselier et al. (1995), comparing the effects of low-temperature storage (5°C or -196°C) on GA, FA, PF bacterioplankton and picophytoplankton cells, found that low but positive storage temperatures resulted in significant and rapid reductions in the total cell count. The study of Wen et al. (2004) has demonstrated the rapid decline in viral numbers over time of viral isolates preserved in aldehyde fixatives (0.5% GA or 2% FA) at 4°C. In their study, viral abundances had decreased by 72% after 16 days. Such results were also confirmed by Ammini et al. (2010) who reported a rapid decline in counts of bacteria and viruses in samples preserved in formaldehyde over a delay of 1 week to 2 months, and they also showed that the decline increased with increase in the final concentration of formaldehyde in the sample. Using liquid nitrogen (blocking all the oxidative reactions responsible for the destruction of organic molecule within cells) and conservation at -80°C provided the best storage conditions on the long-term as already reported elsewhere when samples were fixed (Brussaard 2004). Only a few percentage (<10%) of both cellular and viral loss was indeed recorded between t0 and t30. Our results confirm thus previous studies that reported that preservation of samples at -80°C or in liquid nitrogen (-196°C) is generally much more efficient than preservation at 4°C or -20°C (Brussaard 2004).

In the case of virus counts, the results were somewhat difficult to interpret. When comparing the virus detection efficiency by adding SYBR Green I at a final concentration of 10^{-4} or adding SYBR Gold at a final concentration of 10^{-4}, 5×10^{-5} or 2×10^{-5}, we found that abundances were greatest with SYBR Gold, regardless of the concentration of the stain. The quantitative results were slightly different, if each experiment was considered separately. It appears that at 75°C, SYBR Green I (at a final concentration of 10^{-4}) compared to SYBR Gold (at the three concentrations) incubations, underestimated virus concentrations by 20%, 33% or 35%, respectively. This higher efficiency of SYBR Gold was found at all the incubation temperatures tested (from 45°C to 75°C). Qualitatively, virus populations could be counted more easily when SYBR Gold stained, as the particles yielded higher fluorescence separating them from the low fluorescence background noise of the machine. SYBR Gold also made it possible to distinguish more viral subpopulations if used at low concentrations. The preferential use of low concentrations of SYBR Gold is especially interesting; since SYBR Gold is clearly cheaper than SYBR Green I. Brussaard (2004) tested increasing concentrations of SYBR Gold and SYBR Green I on FCM virus counts, and demonstrated that higher virus counts were obtained by staining the sample with SYBR Green I than with SYBR Gold. She recommended a final concentration of the SYBR Green I stock solution of 5×10^{-5}. However this study was mainly performed using several representatives of different virus families easily distrainable with FCM (and individual populations may react very differently) rather than on natural samples, as in our study. Consistently with her study, we also found that the incubation temperature is very important in order to boost and correctly assess the viral abundance. At low temperatures, there was a significant reduction, and thus a clear underestimation of the total counts.

Taking into account all the dyes and concentrations, we can see that the mean abundances increased significantly by 14% when the temperature was increased from 45°C to 65°C, but there was no significant increase from 65°C to 75°C. It is recognized that heating treatments increase the penetration of the stains by increasing the permeability of the viral capsid and by denaturing the nucleic acids, which may enhance their staining (Marie et al. 1999; Xenopoulos and Bird 1997). Moreover, our study also demonstrated that a higher incubation temperature increased not only the total number of viruses, but also the number of viral subpopulations. It seems that the heat could enhance the "detectability" of different groups of viruses, which might otherwise be invisible because not permeable to the stain. In our study we detected up to 5 different viral populations within the same sample. Analogous to our findings, Chen et al. (2001), revealed the existence of at least four viral subpopulations in a sample from Lake Erie. With regard to the use of dilution solutions, it appeared that no differences were obtained, regardless of whether TE, TBE or filtered lake water was used. Brussard (2004) obtained the highest viral counts when diluting with TE or Tris buffer, and the lowest counts when using distilled water or seawater. We agree that TE is the optimum dilution solution, at least because TE allowed us to differentiate between several subpopulations of viruses. Moreover, regardless of whether the TE buffer had been filtered at 0.02 µm or 0.2 µm, if it was autoclaved, a 30% overestimation of low fluorescence populations, and thus of the total count, could be prevented by removing the background noise in the critical part of the cytogram.

As indicated by Brussaard (2004), over the range tested (between 7 and 8), the pH did not have any influence on the quality of the signature or on the total abundances found. The fixatives tested (GA or FA 1%) did not produce higher final counts of viruses, than unfixed samples. Our findings are consistent with the work done by Wen et al. (2004), which found no significant difference between viral abundance estimates made with fixed (0.5% GA and 2% FA, final concentrations) samples and unfixed samples, provided that the slides were prepared immediately. Brussaard's study (2004) showed, firstly, that there was no significant difference in the FCM signal of fixed and fresh virus samples for fixing lasting up to 1h and, secondly, no conclusive conclusions could be drawn about the use of FA or GA or the best final concentration of the fixative. The two studies indicated above recommended the use of GA at a final concentration of 0.5%, the first one for the reason that occasional reductions in some phytoplankton virus abundances have occurred at higher concentrations. Our study showed that no fixing is necessary if lake water samples are analyzed immediately after their arrival in the laboratory. In addition to this battery of tests, unpublished storage tests performed on some samples stored at 4°C showed that virioplankton abundances had fallen by 40% after being stored for two days. Only fixed and flash-frozen samples can be kept for a long time and delayed analysis.

EFM Counts

At a final concentration of 10^{-3} we found that the fluorescence signal of bacteria and viruses were more stable if they had been SYBR Gold-stained rather than SYBR Green I-stained. These results were in agreement with those presented by Noble (2001) and others indicating that 1) the SYBR Green I signal fades within 30 sec, making it necessary to use an anti-fading solution or higher concentrations of SYBR Green I in order to increase stability

(Noble and Fuhrman 1998) and 2) that the fluorescence of SYBR Gold stained viruses is stable for more than 2 min without any anti-fading solution (Chen et al. 2001). Because of the very fast fading of SYBR Green I, Bettarel et al. (2000) recommended that this stain should not be used for viral concentrations higher than 10^8 particles.mL^{-1}. Wen et al. (2004) reported that the suitability of the two stains depended on the sample being analyzed. From a quantitative point of view, we did not observe any significant increase in bacteria or virus counts depending on whether SYBR Gold or SYBR Green I stain had been used. Neither the type nor the concentration of the fixatives tested (GA 0.5 or 1% and FA 1 or 2%) had any influence on EFM estimations.

Our investigation showed that the slides can be stored at -20°C for up to one month and still provide reliable and realistic counts. From 29 days to 76 days of storage, virus counts fell by 5 and 98% respectively, and bacterial counts by 3 and 73% respectively. Noble (2001) limited the storage of frozen slides to 2-3 weeks, and Wen et al. (2004) reported no decline in viral abundance during the 16 days. Furthermore, Turley and Hughes (1992) reported no decrease in bacterial counts of slides counted immediately after preparation of seawater samples and then after being stored, frozen and recounted within 70 days. It is likely that such differences may be related to the chemical characteristics of seawater or lake water, in conjunction with fixation.

Comparison of FCM and EFM Counts

Our data demonstrated that FCM counts were highly correlated to EFM counts, and that FCM counts were generally higher than EFM ones. Typically, FCM counts of bacteria and viruses were 2.13 and 1.47 times higher respectively than the corresponding EFM counts. Another study, conducted by the end of 2004 within our research laboratory confirmed again this trend with factors of 2.42 and 2.07 for bacteria and viruses, respectively (not shown). The findings of other authors agree with our findings even if the multiplication factors they found were different, *e.g.* 1.1 for Chen et al. (2001) and 1.4 for Marie and coworkers (1999), both obtained on virus samples.

What could contribute to such widely different multiplication factors? We can argue that such differences found between the two techniques can be due to the virus types or bacteria strains, the staining characteristics, etc. A possible explanation of the different factors found for viruses and bacteria could be that viruses are significantly smaller than bacteria, and so the signal is likely to fade very rapidly in EFM. The differences observed between the EFM and FCM counts could also be attributable to a lower accurate estimation by EFM, due to the presence of particles in natural samples, to the fact that fewer cells or biological particles were counted by EFM or to the uneven distribution of the biological entities on the filter (Lebaron et al. 1998). As well as being faster and more accurate, FCM makes it possible to distinguish between different populations. The viral and bacterial dynamics were probably more accurately revealed by FCM, which allowed a finer analysis of shifts in abundances due to its higher resolution.

More generally, it is clear that FCM is less operator-dependent and less labor-intensive than EFM, but that EFM can provide additional information, such as information about the presence of different morphotypes, especially in the case of bacteria. Finally, only EFM permits bacterial phylotyping, as discriminated by the FISH technique (Glöckner et al. 1996), so far.

Conclusion

The major finding of our study is that no unique and universal method exists to assess reliable counts of natural bacterio- and viroplankton populations and that many factors should be considered. It seems however that the confirmation is made that FCM gives better results than EFM and it suggests typically that the concentrations of the total viruses in aquatic ecosystems is probably higher that many previous VLPs estimated found in the literature.

Concerning bacterial counts with FCM, highest numbers were obtained by fixing the samples with GA at a final concentration of 2%, and by incubating the sample at room temperature for 15 min with SYBR Green I at a final concentration of 10^{-4}. From both the quantitative and qualitative considerations, we suggest diluting the samples in TE (0.2- or 0.02-µm filtered). It is important that analyses be done on the same day than sampling, or no more than one day later after conserving the samples at 4°C.

For viral counts using FCM, we recommend using SYBR Gold at a low concentration (2×10^{-5}) in order to obtain high viral counts, and at the same time to be able to access the various subpopulations within the community. For the same reason, the incubation temperature should be 75°C, and the dilution solution recently autoclaved and 0.02 (or 0.2)-µm filtered TE buffer adjusted to a final pH of 7 or 8. Since Marie et al. (1999), FCM analysis has been improved for marine samples by Brussaard et al. (2000), Chen et al. (2001), and Brussaard (2004) proposed the optimized protocol mostly used nowadays for dsDNA viruses on the basis of the analysis of 5 phycoDNAviruses, 1 cyanophage and 6 bacteriophages. In 2007, Tomaru and Nagasaki examined whether this last protocol was also suited for the analysis of small genome viruses with either DNA or RNA genomes using again algal viruses from a marine origin. They could show that the optimum staining protocol may vary significantly among viruses tested. Their article also revealed that for the small viruses (below 40 nm) that harbor either DNA or RNA genomes, FCM counts were similar or lower than values obtained with the Most Probable Number assay, suggesting that only infectious particles could be counted, and more generally speaking that FCM could also sometimes underestimate an important fraction of the smaller viruses in the aquatic plankton. Such a result was also reported recently by Holmfeldt et al. (2011) who revealed that FCM underestimation was clearly obvious for ssDNA. Differently said, the optimum staining protocol may differ among aquatic viruses (i.e. cyanophages, bacteriophages, algal viruses and even in each group, according to the genome type) and all these studies including ours highlight the fact that it may be a non sense to believe that only one protocol exists for all situations. Thus, we have to keep in mind that the protocol we propose here for freshwater viruses is mainly suited for (i) the whole community but that each population/isolate should be tested to find its optimum staining, (ii) dsDNA "large" viruses since it is likely that we miss most of the small containing DNA or RNA particles.

In the case of "our" lake survey, it is important to remind here that we can proceed without fixation that result in considerable eukaryotic cell loss (not shown) and without using liquid nitrogen since we only have a few samples at a time and thus do not need to use extreme temperatures to keep samples. We agree however that both fixation and flash-freezing could be performed as soon as possible after sampling to avoid any evolution of the samples (predation, enzyme activity, and chemical oxidation), particularly if the analyzes cannot be performed the very same day, or last several days or months as usual when hundreds of samples are collected during a cruise or important field survey. Nowadays, dry containers filled with liquid nitrogen, or even Deware's tanks filled with dry ice (CO_2) are available and affordable. As in most sampling cruises hundreds of samples are brought back to the lab, it is obvious that running all of them the very same day is impossible. The need for long term storage and preservation is thus mandatory and this choice must be chosen.

At last, concerning counts done by means of EFM, we recommend that samples should be processed immediately by filtering 1 mL (without dilution), staining the filters with SYBR Gold at a final concentration of 10^{-3} and storing the slides at a temperature of no more than -20°C for up to one month before counting.

Acknowledgment

Ursula Dorigo and Sebastien Personnic were PhD students funded by the French Research Ministry. Monika Ghosh is acknowledged for revising the English version of the manuscript.

References

Ammini, P., Radhakrishnan, S., Sajila, M. P. and Jacob, B. (2007) Study of changes in bacterial and viral abundance in formaldehyde-fixed water samples by epifluorescence microscopy. *Environmental Monitoring and Assessment, 177*, 227-231.

Azam, F., Fenchel, T. Field, J. G., Gray, J. S., Meyer-Reil, L. A. and Thingstad, F. (1983). The ecological role of water-column microbes in the sea. *Marine Ecology Progress Series, 10*, 257-263.

Bergh, O., Borsheim, K. Y., Bratbak, G. and Heldal, M. (1989). High abundance of viruses found in aquatic environments. *Nature, 340*, 467-468.

Bergström, A. K., and Jansson, M. (2000). Bacterioplankton production in humic lake Ortrasket in relation to input of bacterial cells and input of allochthonous organic carbon. *Microbial Ecology, 39*, 101-115.

Berninger, U.-G. (1991). Protozoan control of bacterial abundances in freshwater. *Limnology and Oceanography, 36*, 139-147.

Bettarel, Y., Sime-Ngando, T., Amblard, C., Carrias, J. F., and Portelli, C. (2003). Virioplankton and microbial communities in aquatic systems: a seasonal study in two lakes of differing trophy. *Freshwater Biology, 48*, 810-822.

Bettarel, Y., Sime-Ngando, T., Amblard, C., and Dolan, J. (2004). Viral activity in two contrasting lake ecosystems. *Applied and Environmental Microbiology, 70*, 2941-2951.

Bettarel, Y., Sime-Ngando, T., Amblard, C., Laveran, H. (2000). A comparison of methods for counting viruses in aquatic systems. *Applied and Environmental Microbiology, 66*, 2283-2289.

Bettarel, Y., Sime-Ngando, T., Carrias, J. F., Sargos, D., Garabétian, F., and Lavandier, P. (2003). Viral lysis, flagellate grazing potential, and bacterial production in Lake Pavin. *Microbial Ecology, 45*, 119-127.

Børsheim, K. Y., Bratbak, G., and Heldal, M. (1990). Enumeration and biomass estimation of planktonic bacteria and viruses by transmission electron microscopy. *Applied and Environmental Microbiology, 56*, 352-356.

Bouvier T., del Giorgio, A.P., and Gasol J.M. (2007). A comparative study of the cytometric characteristics of high and low nucleic-acid bacterioplankton cells from different aquatic ecosystems. *Environmental Microbiology,* 9, 2050-2066.

Bratbak, G., and Heldal, M. (1993). Total count of viruses in aquatic environments, p. 135-138. *In* S. Kemp, Sherr, Cole (ed.), Handbook of methods in aquatic Microbial Ecology. Lewis Publisher, Boca Raton Fla.

Brussard, C. P. D. (2004). Optimization of procedures for counting viruses by flow cytometry. *Applied and Environmental Microbiology, 70*, 1506-1513.

Brussard, C. P. D., Marie, D., Bratbak, G. (2000). Flow cytometry detection of viruses. *Journal of Virological Methods, 85*, 175-182.

Buitenhuis, E.T., Li, W. K. W., Lomas, M. W., Karl, D. M., Landry M., Jacquet, S. (2012). Bacterial biomass distribution in the global ocean. *Earth System Science Data,* 5, 301-315.

Chen, F., Lu, J. R., Binder, B. J., Liu, Y. C., and Hodson, E. R. (2001). Application of digital image analysis and flow cytometry to enumerate marine viruses stained with SYBR Gold. *Applied and Environmental Microbiology, 67*, 539-545.

Decamp, O., and Rajendran, N. (1998). Bacterial loss and degradation of bacterial membrane in preserved seawater samples. *Marine pollution bulletin*, 36, 856-859.

Dingman, C. W., and Peacock, A. C. (1968). Analytical studies on nuclear ribonucleic acid using polyacrylamide gel. *Biochemistry, 7*, 659-668.

Domaizon, I., Viboud, S., and Fontvieille, D. (2003). Taxon-specific and seasonal variations in flagellates grazing on heterotrophic bacteria in the oligotrohic lake Annecy - importance of mixotrophy. *FEMS Microbiology Ecology, 46*, 317-329.

Duhamel, S., and Jacquet, S. (2006). Flow cytometric analysis of bacteria and virus-like particles in lake sediments. *Journal of Microbiological Methods, 64*, 316-322.

Fischer, U. R., and Velimirov, B. (2002). High control of bacterial production by viruses in a eutrophic oxbow lake. *Aquatic Microbial Ecology, 27*, 1-12.

Fisher, M. M., Klug, J. L., Lauster, G., Newton, M., and Triplett, E. W. (2000). Effects of resources and trophic interactions on freshwater bacterioplankton diversity. *Microbial Ecology, 40*, 125-138.

Francisco, D. E., Mah, R. A., Taylor, D., and Rabin, A. C. (1973). Acridine orange-epifluorescence technique for counting bacteria in natural waters. *Trans American Microscopy Society, 92*, 416-421.

Fuhrman, J. A. (1999). Marine viruses and their biogeochemical and ecological effects. *Nature, 399*, 541-547.

Gasol, J. M., Zweifel, U. L., Peters, F., Fuhrmann, J. A., and Hagström, A. (1999). Significance of size and nucleic acid content heterogeneity as measured by flow cytometry in natural planktonic bacteria. *Applied and Environmental Microbiolology, 65*, 4475-4483.

Glöckner, F. O., Amann, R., Alfreider, A., Pernthaler, J., Psenner, R., Karlheinz, T. and Schleifer, K. H. (1996). An *in situ* hybridization protocol for detection and identification of planktonic bacteria. System. *Applied Microbiology, 19*, 403-406.

Gundersen, K., Bratbak, G., and Heldal, M. (1996). Factors influencing the loss of bacteria in preserved seawater samples. *Marine Ecology Progress Series, 137*, 305-310.

Hall, J. (1991). Long-term preservation of picoplankton for counting by fluorescence microscopy. *British Phycological Journal*, 26, 169-174.

Hara, S., Terauchi, K., and Koike, I. (1991). Abundance of viruses in marine waters: assessment by epifluorescence and transmission electron microscopy. *Applied and Environmental Microbiolology, 57*, 2731-2734.

Hayat, M. A. (1970). Principles and techniques of electron microscopy. Biological Applications. Van Nostrand Reinhold CO., New York, Vol. 1.226.

Holmfeldt, K., Odic, D., Sullivan, M. B., Middelboe, M., Riemann, L. (2011). Cultivated single-stranded DNA phages that infect marine Bacteroidetes prove difficult to detect with DNA-binding stains. *Applied and Environmental Microbiolology, 78*, 892-894.

Hennes, K. P., and Suttle, C. A. (1995). Direct counts of viruses in natural waters and laboratory cultures by epifluorescence microscopy. *Limnology and Oceanography, 40*, 1050-1055.

Jacquet, S., Lennon, J. F., and Vaulot, D. (1998). Application of a compact automatic sea water sampler to high frequency picoplankton studies. *Aquatic Microbial Ecology, 14*, 309-314.

Jacquet, S., Domaizon, I., Personnic, S., Duhamel, S., Pradeep Ram, A.S., Heldal, M., Sime-Ngando, T. (2005). Estimates of protozoan and virus-mediated mortality of bacterioplankton in Lake Bourget (France). *Freshwater Biology*, 50, 627-645.

Jacquet, S., Miki, T., Noble, R., Peduzzi, P., and Wilhelm, S. (2010). Viruses in aquatic ecosystems: important advancements of the last 20 years and prospects for the future in microbial oceanography and limnology. *Advances in Oceanography and Limnology, 1*, 71-101.

Jochem, F. J. (2001). Morphology and DNA content of bacterioplankton in the northern Gulf of Mexico: analysis by epifluorescence microscopy and flow cytometry. *Aquatic Microbial Ecology, 25*, 179-194.

Kiernan, J.A. (2000). Formaldehyde, formalin, paraformaldehyde and glutaraldehyde: What they are and what they do? Microscopy Today, 8-12.

Larsen, A., Fonnes Flaten, G. A., Sandaa, R. A., Castberg, T., Thyrhaug, R., Rune Erga, S., Jacquet, S., and Bratbak, G. (2004). Spring phytoplankton bloom dynamics in Norwegian coatal waters: microbial community succession and diversity. *Limnology and Oceanography, 49*, 180-190.

Lebaron, P., Catala, P., and Parthuisot, N. (1998). Effectiveness of SYTOX green stain for bacterial viability assessment. *Applied and Environmental Microbiolology, 64*, 2697-2700.

Lebaron, P., Parthuisot, N. and Catala, P. (1998). Comparison of blue nucleic acid dyes for flow cytometric enumeration of bacteria in aquatic systems. *Applied and Environmental Microbiolology, 64*, 1725-1730.

Lebaron, P., Servais, P. Baudoux, A. C., Bourrain, M. Courties, C. and Parthuisot, N. (2002). Variations of bacterial-specific activity with cell size and nucleic acid content assessed by flow cytometry. *Aquatic Microbial Ecology, 28*, 131-140.

Lemarchand, K., Parthuisot, N., Catala, P., and Lebaron, P. (2001). Comparative assessment of epifluorescence microscopy, flow cytometry and solid-phase cytometry used in the enumeration of specific bacteria in water. *Aquatic Microbial Ecology, 25*, 301-309.

Li, W. K. W., and Dickie, P. M. (2001). Monitoring phytoplankton, bacterioplankton, and virioplankton in a coastal inlet (Bedford Basin) by flow cytometry. *Cytometry, 44*, 236-246.

Lisle, J. T., Hamilton, W. A., Willse, A. R., and McFeters, G. A. (2004). Comparison of fluorescence microscopy and solid-phase cytometry methods for counting bacteria in water. *Applied and Environmental Microbiolology, 70*, 5343-5348.

Marie, D., Brussard, C. P. D., Thyrhaug, R., Bratbak, G., and Vaulot, D. (1999). Enumeration of marine viruses in culture and natural samples by flow cytometry. *Applied and Environmental Microbiolology, 65*, 45-52.

Marie, D., Partensky, F., Jacquet, S., and Vaulot, D. (1997). Enumeration and cell cycle analysis of natural populations of marine picoplankton by flow cytometry using the nucleic acid stain SYBR Green I. *Applied and Environmental Microbiolology, 63*, 186-193.

Marie, D., Partensky, F., Vaulot, D., and Brussaard, C. P. D. (1999). Enumeration of phytoplankton, bacteria and viruses in marine samples, p. 11.11.11-11.11.15., Current Protocols in Cytometry. John Wiley and Sons, Inc, New York.

Middelboe, M., and Glud, R. N. (2003). Distribution of viruses and bacteria in relation to diagenetic activity in an estuarine sediment. *Limnology and Oceanography, 48*, 1447-1456.

Navaluna, N. A., Perry, M. J., Talbot, M. C. (1989). The effects of chemical fixation on some optical properties of phytoplankton. *Journal of Plankton Research, 11*, 15-24.

Noble, R. T. (2001). Enumeration of viruses, p. 43-51. In J. H. Paul (ed.), Marine Microbiology, Methods in Microbiology, vol. 30. Academic Press, NY.

Noble, R. T., and Fuhrman, J. A. (1998). Use of SYBR Green I for rapid epifluorescence counts of marine viruses and bacteria. *Aquatic Microbial Ecology, 14*, 113-118.

Ogden, R., and Deborah, A. (1987). Guide to molecular cloning techniques, p. 75-78. In S. L. a. A. R. K. e. Berger (ed.), Methods in Enzymology. Academic Press Inc., Orlando (Florida).

Øvreås, L., Bourne, D. A., Sandaa, R. A., Casamayor, E. O., Benlloch, S., Goddard,V. J., Smerdon, G., Heldal, M., and Thingstad, T. F. (2003). Response of bacterial and viral communities to nutrient manipulations in seawater mesocosms. *Aquatic Microbial Ecology, 31*, 109-121.

Paul, M. L., Jiang, S. C., and Brose, J. B. (1991). Concentration of viruses and dissolved DNA from aquatic environments by vortex flow filtration. *Applied and Environmental Microbiolology, 57*, 2197-2204.

Personnic, S., Domaizon, I., Dorigo, U. Berdjeb, L., and Jacquet, S. (2009). Seasonal and spatial variability of virio, bacterio- and picophytoplanktonic abundances in three peri-alpine lakes. *Hydrobiologia, 627*, 99-111.

Proctor, L. M., and Fuhrman, J. A. (1990). Viral mortality of marine bacteria and cyanobacteria. *Nature, 343*, 60-62.

Schröder, D. C., Oke, J., Hall, M., Malin, G. and Wilson, W. H. (2003). Virus succession observed during an *Emiliania huxleyi* bloom. *Applied and Environmental Microbiolology, 69*, 2484-2490.

Shopov, A., Williams, S. C., and Verity, P. G. (2000). Improvements in image analysis and fluorescence microscopy to discriminate and enumerate bacteria and viruses in aquatic samples. *Aquatic Microbial Ecology, 22*, 103-110.

Simek, K., Macek, K., and Vyhnalek, V. (1990). Uptake of bacteria-sized fluorescent particles by natural protozoan assemblage in a reservoir. *Archives of Hybrobiologia Beih, 34*, 275-281.

Simek, K., Pernthaler, J., Weinbauer, M. G., Hornak, K., Dolan, J. R., Nedoma, J., Mašín, M., and Amann, R. I. (2001). Changes in bacterial community composition and dynamics and viral mortality rates associated with enhanced flagellated grazing in a mesoeutrophic reservoir. *Applied and Environmental Microbiolology, 67*, 2723-2733.

Sime-Ngando, T. (1997). Importance des virus dans la structure et le fonctionnement des réseaux trophiques microbiens aquatiques. *Annales de Biologie, 36*, 181-210.

Sime-Ngando, T., Bettarel, Y., Chartogne, C., and Sean, K. (2003). The imprint of wild viruses on freshwater microbial ecology. *Recent Research and Development in Microbiology, 7*, 481-497.

Tomaru, Y., and Nagasaki, K. (2007) Flow cytometric detection and enumeration of DNA and RNA viruses infecting marine eukaryotic microalgae. *Journal of Oceanography, 63*, 215-221.

Troussellier, M., Courties, C. and Zettelmaier, S. (1995). Flow cytometric analysis of coastal lagoon bacterioplankton and picophytoplankton: fixation and storage effects. *Estuarine, Coastal and Shelf Science, 40*, 621-633.

Turley, C. M., and Hughes, D. J. (1992). Effects of storage on direct estimates of bacterial numbers of preserved seawater samples. *Deep-Sea Research, 39*, 375-394.

Turley, C.M., and Hughes, D.J. (1994). The effect of storage temperature on the enumeration of epifluorescence-detectable bacterial cells in preserved seawater samples. *Journal of the marine biological association* U.K., 74, 259-262.

Vaulot, D., Courties, C., and Partensky, F. (1989). A simple method to preserve oceanic phytoplankton for flow cytometric analyses. *Cytometry, 10*, 629-635.

Vives-Rego, J., Lebaron, P., and Neve-von Caron, G. (2000). Current and future applications of flow cytometry in aquatic microbiology. *FEMS Microbiology Ecology, 24*, 429-448.

Watson, S. W., Novitsky, T. J., Quinby, H. L., and Valois, F. W. (1977). Determination of bacterial number and biomass in the marine environment. *Applied and Environmental Microbiolology, 33*, 940-946.

Weinbauer, M. G., and Rassoulzadegan, F. (2004). Minireview: Are viruses driving microbial diversification and diversity? *Environmental Microbiology, 6*, 1-11.

Wen, K., Ortmann, A. C., and Suttle, C. A. (2004). Accurate estimation of viral abundance by epifluorescence. *Applied and Environmental Microbiolology, 70*, 3862-3867.

West, C. A., He, C., Su, M., Swanson, S. J., and Mentzer, S. J. (2001). Aldehyde fixation and thiol-reactive fluorescent cytoplasmic probes for tracking cell migration. *Journal of Histochemistry and Cytochemistry, 49*, 511-518.

Wilhelm, S. W., and Suttle, C. A. (1999). Viruses and nutrient cycles in the sea. *BioScience, 49*, 781-787.

Wommack, K. E., and Colwell, R. R. (2000). Virioplankton: viruses in aquatic ecosystems. *Microbiology and Molecular Biology Review, 64*, 69-114.

Xenopoulos, M. A., and Bird, D. F. (1997). Virus à la sauce Yo-Pro: Microwave-enhanced staining for counting viruses by epifluorescence microscopy. *Limnology and Oceanography, 42*, 1648-1650.

In: Flow Cytometry
Editor: Stefanos Papandreou
ISBN: 978-1-62808-709-3

Chapter 2

Application of Flow Cytometry in Phenotyping, Analysis and Functional Characterization of T Regulatory and Immune Suppressor Cells: A Focus on Multiple Myeloma Immune Regulatory Biology

***Karthick Raja Muthu Raja*[1,*] *and Roman Hajek*[1,2,3,4]**
[1]Babak Myeloma Group, Department of Pathological Physiology, Faculty of Medicine, Brno, Czech Republic
[2]Department of Hemato-oncology, University Hospital of Ostrava, Ostrava, Czech Republic
[3]Department of Clinical Studies, University of Ostrava, Ostrava, Czech Republic
[4]Department of Clinical Hematology, University Hospital of Brno, Brno, Czech Republic

Abstract

Flow cytometry is an inevitable tool in clinical diagnosis of various pathological conditions. Recently, its application plays a significant role in monitoring of several immune suppressor cells in patients with cancer and autoimmune disorders. Regulatory T cells (Treg cells) are immune cells which help in maintaining immune homeostasis in physiological conditions by patrolling unwanted immune responses; however, in cancers and autoimmune diseases, Treg cell numbers and function are significantly altered. In this chapter, we focus and update the information regarding Treg cells and their subsets, and

[*] E-mail: rajakarthick594@gmail.com; Phone: +420549498450; Fax: +420549498480.

application of flow cytometry in phenotyping and functional characterization of various immune regulatory and suppressor cells in multiple myeloma patients.

Introduction

T regulatory (Treg) cells were formerly known as T suppressor cells [1, 2]. Earlier studies considered that a subset of CD8 T cells possessing suppressive activity as T suppressor cells. For the first time, North's group showed the presence of CD4 T cells with suppressive function that abrogated anti-tumor response in murine models [3]. Subsequently, understanding of peripheral CD4 T cell mediated immune regulation in animal models leads to *in vitro* human tumor model studies. In a human melanoma model, isolated CD4+ Treg cells/suppressor cells from blood and tumor tissues were able to suppress the generation of cytotoxic T cell responses [4]. With the acknowledgement of *in vitro* and mice tumor model studies regarding Treg cells, an advancement happened in early 2000's to study the Treg cells and their suppressive function in various cancers. Several groups reported that CD4 Treg cells from cancer patients were increased and freshly isolated Treg cells were able to abrogate the effector T cells. Therefore, most of groups believed that increased Treg cells could have negative impact on anti-tumor response [5]. Before going in detail, we would like to summarize ontogeny, significance of FoxP3 and suppression mechanisms of Treg cells.

Ontogeny of T Regulatory Cells

Evidence of thymic origin of T regulatory (Treg) cells and their presence in the periphery was clearly shown in T-cell deficient mice by adoptively transferring thymocytes or T cells depleted of Treg cells which resulted in autoimmune diseases [1, 6, 7]. Treg cells are detectable approximately in a period of 3 days after birth. Studies have clearly shown that neonatal thymectomy leads to abrogation of Treg cells in the thymus [6, 8]. In mice, even depletion of Treg cells for a short period leads to autoimmunity [8, 9]. These above mentioned findings suggest that thymus derived Treg cells are natural Treg cells. Presence of immature Treg cells was observed in human thymus but information regarding the microenvironment for thymic generation of Treg cells is insufficient [10, 11]. In mice models, several stromal cells (including thymic cortical cells, thymic medullary cells and dendritic cells) have been reported for thymic selection and differentiation of Treg cells [12]. Thymic development of Treg cells requires high affinity T cell receptor interaction with stromal cell self-peptides presented by MHC molecule [13]. Apart from self-peptide presentation, these stromal cells also provide co-stimulatory signals (CD28 or CD40) for the Treg cell development [12]. Soluble factors, such as IL-2 and IL-7 cytokines, are required for thymic generation of Treg cells [14]. Generation of Treg cells in the thymus is more similar in mice and humans but unique structure (Hassall's corpuscles) was reported in generation of human Treg cells. This Hassall's corpuscles in thymus provide a specialized environment for nascent Treg cells [15]. Hassall's corpuscles secrete a protein called thymic stromal lymphopoietin (TSLP) which activates thymic dendritic cells (DCs) [15, 16]; these activated DCs mediate expression of forkhead box P3 (FoxP3) in immature Treg cells but not in mature naïve Treg cells from peripheral blood (PB) [17]. TSLP activated DCs enhance high affinity generation of thymic T cells and differentiation into mature Treg cells [18].

Ontogeny of Treg cells generation in mice and humans might be similar, but in humans most of the T cells development occurs *in utero*. Mature T cells can be detected in the early 13th week of gestation; by 14th week approximately 7% of mature T cells start to express CD25 which is a marker of Treg cells [19, 20]. Studies have shown that thymic Treg cells isolated from fetus suppressed effector T cells in co-culture experiments and expressed FoxP3 molecule [20]. These data suggest that human Treg cells begin to develop *in utero* itself.

FoxP3 Gene and Their Role in T Regulatory Cell Differentiation

FoxP3 gene contains 11 coding exons and is located on the *p* arm of the X chromosome. In the last decade, studies have shown that loss of FoxP3 function and mutation caused severe multi-organ autoimmune and inflammatory disorder called immunodysregulation polyendocrinopathy enteropathy X-linked syndrome (IPEX) in humans. Similarly, FoxP3 mutant mouse strain had devastating lesions [21, 22, 23]. In humans and mice, FoxP3 mutant males were only affected but not heterozygote females. With the cause of immune mediated lesions, FoxP3 mutation was also consistent [9, 24]. Several studies reported that FoxP3 expression was highly expressed in CD25+ cells than in others. Experimental data regarding retroviral vector mediated expression of FoxP3 gene in CD25- T cells have shown that transformed cells possessed suppressive function and phenotype of Treg cells [9, 24]. Also, FoxP3 significance was clearly demonstrated in chimeric mice, which were transferred with mixed bone marrow (BM) cells from FoxP3 mutant and wild type mice. These transformed mice were functional similarly to healthy mice; also, these chimeric mice experiments showed that Treg cells originated only from FoxP3 wild type strain but not from mutant mice [9]. These findings clearly suggest that FoxP3 expression is a substantial factor for differentiation of Treg cells.

Role of FoxP3 in T Regulatory Cells

Humans express two isoforms of FoxP3, either of one can confer suppressive function when strongly expressed [25, 26]. The deletional FoxP3 isoform (FoxP3B) lacks proline-rich exon and absence of amino-terminal residues prevent the interaction with nuclear factor of activated T cells. Therefore, no real suppression can be seen [27, 28]. Role of FoxP3B isoform in human Treg cells remains elusive. Transient expression of FoxP3 does not confer suppressive activity of Treg cells [25, 29]. The suppressive ability of Treg cells is highly dependent on high and stable expression of FoxP3. Studies have shown that long term culturing of Treg cells affected their suppressive activity [30, 31]. In humans, naive CD4 T cells can be converted (*in vitro*) into FoxP3+ Treg cells by TCR stimulation and TGF-β. These induced Treg cells do not possess similar suppressive function of natural Treg cells; also, they secrete pro-inflammatory cytokines [29, 32]. FoxP3 is expressed in both activated and regulatory T cells and also share similar phenotypic antigens, including CD25, cytotoxic

T-lymphocyte antigen (CTLA-4), glucocorticoid induced tumor necrosis factor (GITR) and CD95 [33]. It was proposed that expression of FoxP3 in activated T cells as one of the homeostasis mechanisms which exert negative feedback effect during the course of an immune response [33].

Suppressive Mechanisms of T Regulatory Cells

T Regulatory Cell Mediated Suppression of T Cells

Several *in vitro* and *in vivo* analyses have shown that Treg cells suppress proliferation and inhibition of cytokine secretion by effector T cells. These Treg cells also inhibit conversion of CD8 T cells into cytolytic effector cells without influencing their proliferation or cytokine secretion [34]. Treg cells can target effector T cells independent of antigen presenting cells (APCs) [35]. This observation suggests contact-dependent mechanism of Treg cell mediated suppression. Release of cyclic adenosine phosphate (cAMP) by Treg cells into effector T cells via gap junctions also suppresses their proliferation and function [36]. Some studies have shown that perforin and granzyme B molecules were involved in Treg cell mediated direct targeting of effector T cells [37, 38, 39]. Apart from the release of suppressive molecules in direct killing, these Treg cells also induce apoptosis in effector T cells via Bim molecule, a pro-apoptotic protein. Bim molecule that mediates apoptosis was found to be reversed in the presence of anti-apoptotic molecule, Bcl-2 protein [40].

T Regulatory Cells Mediated Suppression of Antigen-Presenting Cells

Interaction of Treg cells with DCs affects the engagement and activation of effector T cells by the same DCs. This interaction leads to abrogation of DCs antigen presentation and enhances secretion of suppressive molecules. It has been documented that activation of DCs with Treg cells up regulates the indoleamine 2,3-dioxygenase (IDO), a suppressive enzyme associated with tumor immune evasion and pregnancy [41, 42]. Regarding IDO, findings have shown that IDO induction depended on high CTLA-4 expression on Treg cells [43]. *In vitro* experimental data have shown that blockade of CTLA-4 and CTLA-4 deficient mice models did not show abrogation of Treg cell suppressive function [44, 45]. However, *in vivo* blockade of CTLA-4 abrogates the Treg cells in inflammatory bowel disease model. Findings have suggested that CTLA-4 deficient Treg cells suppress via IL-10 and TGF-β *in vivo* and *in vitro*, respectively [44, 46]. In mice models, it was found that FoxP3 and CTLA-4 molecules complement each other in suppressive function of Treg cells [47]. Findings regarding Treg cell mediated contact-dependent suppression suggest that Treg cells do not rely on one mechanism alternatively on various other mechanisms to control unwanted immune responses.

T Regulatory Cell Derived Soluble Suppressor Molecules

In vitro and *in vivo* studies claim opposing results regarding the IL-10 and TGF-β mediated suppressive effects. *In vitro* studies have shown that after neutralizing IL-10 and TGF-β cytokines with monoclonal antibodies suppressive function of Treg cells was not abolished. In contrast, *in vivo* studies (mouse models of inflammatory bowel disease and type1 diabetes) demonstrated that Treg cell function was abrogated after neutralizing the IL-10 and TGF-β cytokines [40]. TGF-β mediates expression of FoxP3 in conventional T cells [48]. IL-10 from Treg cells abrogates the DC function and induces generation of Tr1 regulatory cells from conventional T cells [49]. HO-1 molecule was found to be linked with Treg cell function, which catalyzes the formation of carbon monoxide through heme degradation [50, 51]. Studies have proposed that Treg cells can suppress T cells via secreting IL-35 cytokine, which is composed of IL-12 p35 chain and Ebi3, and regulated by FoxP3. Abrogation of either IL-12 p35 chain or Ebi3 molecule reduces the function of Treg cells *in vitro* and unable to control effector T cell induced inflammatory bowel disease in mice models [52].

Delineation of Phenotypic Features of CD4 T Regulatory Cells

In humans, characterization of Treg cells remains fuzzy. Using FoxP3, one can identify Treg cells. However, for functional studies, it is impossible to use FoxP3 as an identifying marker. So, it is obvious that isolation of Treg cells for functional studies requires other surface markers. One of the well-known markers is CD25 but nearly 30% of T cells do express CD25 from that approximately about 1-2% of T cells with high expression of CD25 possess suppressive activity [53, 54]. Several reports suggested that CD25 and FoxP3 expression was proportional. However, it is noteworthy that selection of Treg cells with high CD25 expression excludes naïve Treg cells with low FoxP3 and moderate CD25 expression [33]. Therefore, CD25 gating boundaries between laboratories vary especially for isolation of Treg cells and thus, influence reproducibility of results. Several groups used CD127 marker for isolation of Treg cells because these cells express low level of CD127 [55, 56]. However, one may not solely depend on this marker due to its down regulation also in activated T cells [57, 58]. Some studies suggested that CD62L molecule can be used to differentiate Treg cells (62L high) from recently activated T cells (62L low) [59]. Combined use of CD25 and CD127 markers to isolate Treg cells found to be contaminated by non-Treg cells expressing FoxP3. These non-Treg cells express memory marker CD45RO and FoxP3 but lack the suppressive function, and they are able to secrete pro-inflammatory cytokines [60]. Main feature of FoxP3 in these non-Treg cells is incomplete demethylation of FoxP3, whereas FoxP3 in Treg cells with suppressive function is completely demethylated [33]. These above mentioned findings suggest that presence of non-Treg cells with low expression of FoxP3 excludes FoxP3 as an independent marker of Treg cells. Figure 1 summarizes the classical phenotypisation of Treg cells. Molecules associated with Treg cells in terms of origin, homing, activation and function are summarized in Table 1.

Subsets of CD4 T Regulatory Cells

Apart from naturally occurring or thymus originated Treg cells, there are several inducible Treg cells are present. Mainly inducible Treg cells are formed in the periphery from naive T cells in the presence of several soluble factors, such as IL-10, IL-35 and TGF-β. It has been proposed that inducible Treg cells do not express FoxP3 compared to natural Treg cells [61]. Natural and inducible subsets of CD4 Treg cells are summarized in Table 2.

Table 1. Summary of molecules associated with T regulatory cell origin, activation and function

Transcription factor	FoxP3
Homing and origin	CD62L, CCR4, CCR6, CCR9, CD103, CD304, CD31, Lack of CD49d
Activation and memory	CD45RA, CD45RO, CD25, HLA-DR, Lack of CD127, CD69
Suppressive function	CTLA4, ICOS, CD39–CD73, LAP, Granzyme B, Galectin 1, Galectin 10, TRANCE, CD80 and CD86, IL-10, IL-17, CD2, Lack of IL-2
Apoptosis and survival	CD27, OX40, CD95, PD1, GITR, Galectin 3, GARP, MS4A4B, IL-1R, CD6

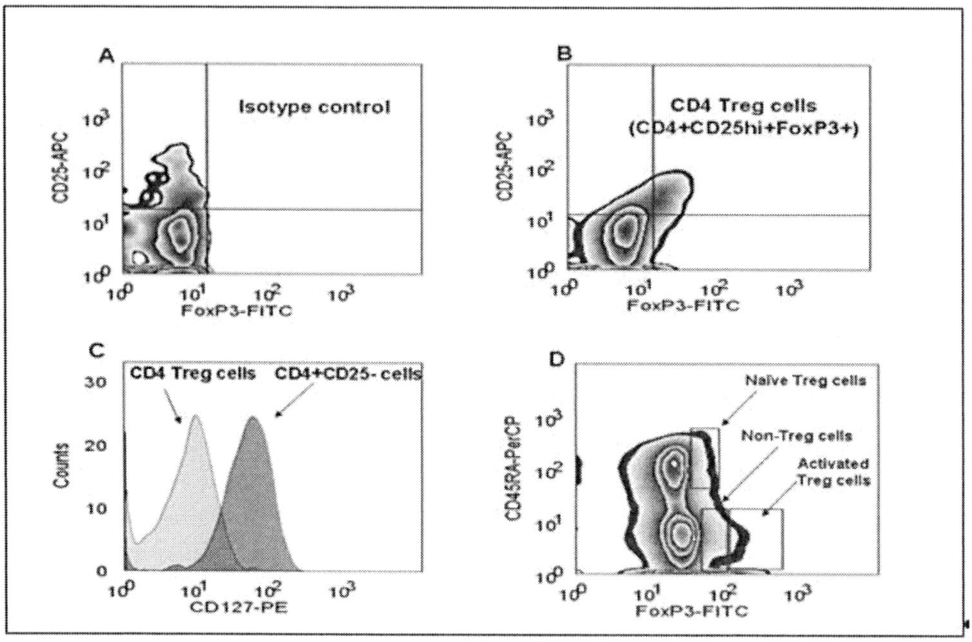

Figure 1. Phenotypisation of CD4 T regulatory cells and their subsets. As shown in dot plot (B), CD4 Treg cells are identified by a high expression of IL-2 receptor (CD25) and FoxP3 (intracellular). CD4 Treg cells do not express or very dimly express CD127 compared to effector T cells (C). CD4 Treg cell subsets are characterized using markers CD45RA and FoxP3. Naïve, non-Treg and activated Treg cells are identified as: CD4+CD45RA+FoxP3dim+, CD4+CD45RA-FoxP3dim+ and CD4+CD45RA-FoxP3hi+, respectively.

Assessment of Suppressive Function of T Regulatory Cells: Flow Cytometry Versus Thymidine Uptake Method

Unfortunately, until present there are no *in vivo* functional data to appreciate the Treg cell mediated suppression. There is no evidence that *in vitro* functional assay reflects the *in vivo* function of Treg cells [33]. Generally, most of the research group assumes that Treg cells are non-proliferative and prefers to perform thymidine uptake method due to the fact that mouse Treg cells are unresponsive *in vitro* culture condition.

Table 2. Natural and inducible subsets of CD4 T regulatory cells

Types	Origin	Functions	References
CD4+ natural Treg cells	Arise in thymus and disseminate to periphery; constitute about 10%-15% of CD4 cells	In mice, depletion of these cells leads to autoimmunity. In humans, mutation of FoxP3 gene, which is located on X chromosome leads to fatal immune disorder IPEX syndrome	[62]
CD4+ Tr1 regulatory cells	Periphery	Induced in the periphery from naïve T cells in the presence of IL-10. These cells lack FoxP3 expression but secrete IL-10 and TGF-β.	[63, 64]
CD4+ Th3 cells	Periphery	Induced in the periphery from naïve T cells in the presence of TGF-β. Suppression is mediated by TGF-β. Rare Th3 cells express FoxP3 molecule due to induction by TGF-β.	[48, 65]
Double negative Treg cells	Periphery	In mice and humans these cells constitute about 1%-3% and 1%, respectively. Double negative Treg cells inhibit T cell activation and proliferation in an antigen specific manner.	[66]
γδ T cells	Infiltrate into tumor site (breast cancer)	Suppress naïve and effector T cell responses, and inhibit maturation and function of dendritic cells.	[67]
iTR35 (IL-35 induced Treg cells)	Inducible from conventional T cells	Independent of FoxP3 expression. Strongly suppressive and highly stable *in vivo*. *In vitro* iTR35 could be generated by IL-35 and IL-10 treatment. Inducible in inflammatory conditions and tumor microenvironment.	[52]

In thymidine uptake method, one can assess the proliferating cells which uptakes thymidine. In humans, it has been shown that naïve Treg cells (CD45RA+) proliferate actively into effector Treg cells (CD45RA-) in *in vitro* co-culture assays. In this case, thymidine uptake method might not be suitable to assess the Treg cell suppressive function. To overcome this issue, we propose that dilution of 5, 6-carboxyfluorescein diacetate succinimidyl ester (CFSE) dye in effector T cells can be used to assess the suppressive status of Treg cells via indirectly measuring the proliferation of CFSE labeled effector T cells in flow cytometry [68, 69] (Figure 2).

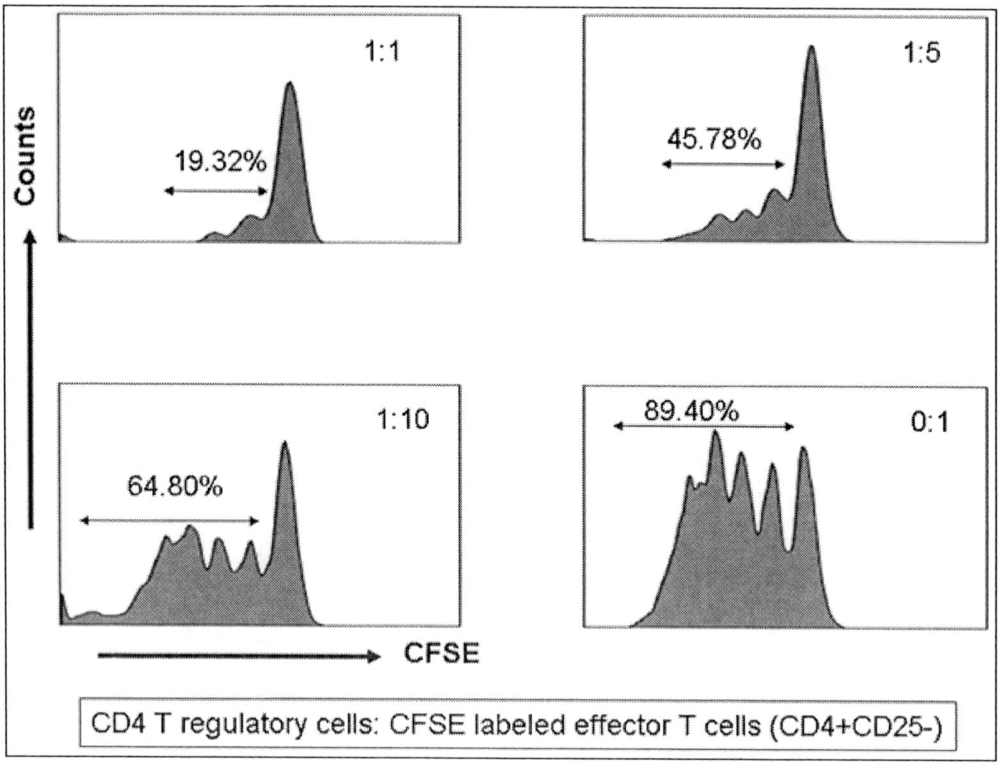

Figure 2. CFSE based proliferation/suppression assay for the assessment of function of T regulatory cells. This figure represents suppressive activity of Treg cells isolated from a myeloma patient. Isolated CD4 effector T cells were CFSE labeled and stimulated with anti-CD3/CD28 beads in a 1:1 ratio, and co-cultured with different concentrations of Treg cells as mentioned above in the figure. As feeder cells, mitomycin-C treated non-T cells were used. After 4 days of cultivation, cells were collected and analyzed in flow cytometry for the dilution of CFSE (FL-1 channel) from effector T cells. The above figure clearly explains Treg cell mediated suppression of effector T cells in a concentration depend manner, which is clearly shown by dilution of the CFSE dye.

CD8 T Regulatory Cells: A Novel Suppressor

T cells play an important role in immunosurveillance against cancers. Eventually, these T cells may become CD4/CD8 regulatory T cells due to stimulation and interaction with tumor cells. Thus, these generated regulatory cells promote the growth of tumor rather than inhibition of tumor [70]. Mechanisms of suppression and expansion are well documented for CD4 Treg cells but recent research has been directed to screen the presence of CD8 Treg cells in various tumors and inflammatory conditions. Recent reports suggest that CD8 Treg cells suppress an immune response by several mechanisms including cell-cell contact, non-specific cytokines and specific killing mechanisms [71, 72, 73]. CD8 Treg cells can be categorized into two groups, on the basis of antigen specificity (Table 3).

Table 3. Classification of CD8 T regulatory cells according to antigen specificity

Specificity	Phenotypes of CD8 Treg cells
Antigen specific	CD8+CD28-, CD8+CD75s+, CD8+CD45RChi TC1 and TCR peptide specific CD8αα Treg cells
Non-antigen specific	CD8+CD25+, CD8+CD122+ and CD8+CD45RClow Treg cells

Suppressive Mechanisms of CD8 Treg Cell Subsets

CD8+CD28- Treg Cells

These Treg cells are alloantigen specific Treg cells. They impose suppression via cell-cell contact mechanism [74, 75]. These CD8 Treg cells interfere with antigen presenting cells (APCs) and non-professional APCs via NF-kB inactivation and maturation of DCs [76]. Thereby, interaction of tolerogenic DCs with CD4 T cells leads to regulation of CD4 T cells. This type of mechanism has been reported in heart transplanted patients [76]. Transcription factor FoxP3 is expressed in its different isoform (FoxP3α) due to lacking of exon3 region [70].

Naturally Occurring CD8 Treg Cells

CD8+CD25hi+ and CD8+CD45RClow Treg cells reported to have similar phenotypic features of CD4 Treg cells (FoxP3, CTLA-4, GITR and CD127) (Figure 3) [73, 77, 78]. These cells were found to suppress the CD4 T cells in a contact dependent mechanism. Presence of increased numbers of naturally occurring CD8 Treg cells has been recently documented in several cancers including prostate [79], nasopharyngeal [80], colorectal [77] and multiple myeloma (MM) [78]. Interestingly, a study has suggested that CD8+CD25hi+ Treg cell clone isolated from thymocytes did not impede Th2 cells compared to Th1 cells [81]. This study has also reported that Th2 cells are refractory to CD8+CD25hi+ Treg cells in an IL-4 dependent manner [81]. CD8+CD45RClow Treg cells were reported in animal model studies and found to protect against CD4 T cell mediated graft versus host disease [82].

Qa-1 Restricted CD8 Treg Cells

In autoimmune disease models, such as experimental allergic encephalomyelitis (EAE) and systemic lupus erythematosus (SLE) suppressive activity of a novel CD8 Treg cell has been documented recently [83]. These CD8 Treg cells target cells expressing MHC Qa-1 in murine models; in humans, this molecule is termed as HLA-E. Dimerization of Qa-1 with β2M molecule of lymphoid cells induces presentation of self or foreign antigens. Thereby, T and B cells expressing Qa-1- β2M heterodimeric molecule turned into a target for a subtype of CD8 T cells [83]. In a study, it was shown that disruption of interaction between Qa-1 restricted CD8 Treg cells and CD4 T cells induces SLE like autoimmune disease in murine models [84].

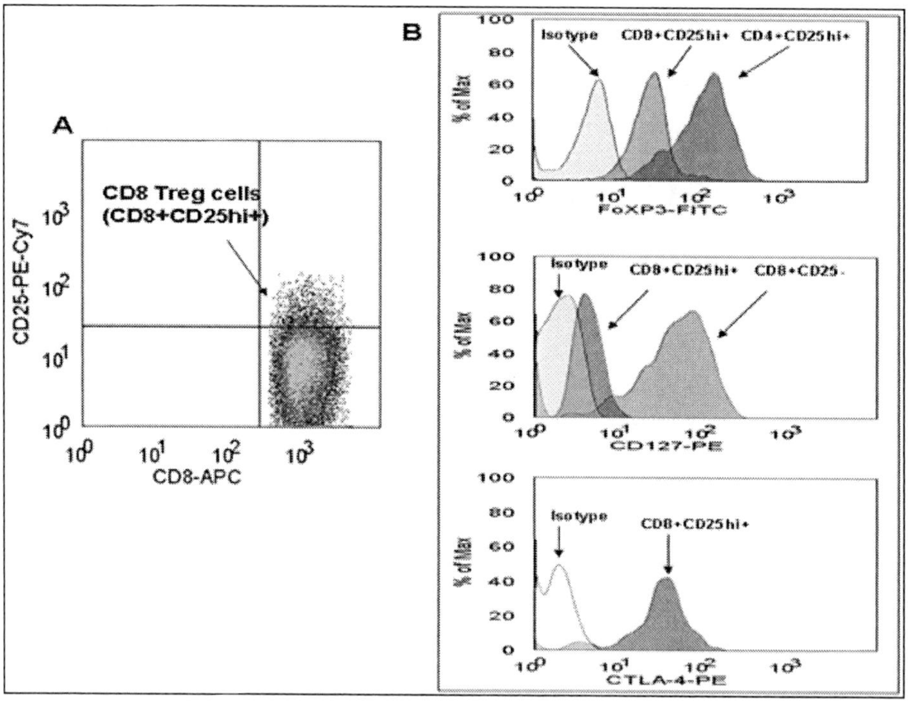

Figure 3. Phenotypisation of CD8 T regulatory cells. As shown in dotplot (A), one of the naturally occurring CD8 Treg cells are identified based on a high expression of IL-2 receptor (CD25) on CD8 T cells. Further, these cells are characterized for expression of regulatory phenotype markers, including FoxP3 (intracellular), CD127 and CTLA-4. Similarly to CD4 Treg cells, CD8 Treg cells express FoxP3 (but not with high intensity), CD127dim/- and CTLA-4 (B).

Myeloid-Derived Suppressor Cells

Myeloid-derived suppressor cells (MDSCs) are activated immature myeloid cells (IMC) that have been prevented from differentiation to mature cells. These cells are expanded in pathological conditions [85]. MDSCs lack the expression of cell surface markers that are specifically expressed by monocytes, macrophages or DCs, and comprise a mixture of myeloid cells that have the morphology of granulocytes or monocytes [86]. They are potent suppressors of T cells. In cancer patients and tumor models, accumulation of MDSCs occurs due to release of soluble factors by tumor cells or cells in the tumor environment [87, 88].

Phenotypic Features of Myeloid-Derived Suppressor Cells

Human MDSCs can be characterized phenotypically by CD33 expression which is a common marker for myeloid cells. These suppressor cells strictly lack lineage antigens, such as CD3, CD16, CD19, CD20 and CD56. MDSCs can be divided into two subsets: monocytic (CD14) and granulocytic MDSCs (CD15), and both of these subsets express CD33. Moreover, MDSCs lack the expression of mature lymphoid and myeloid markers as well MHC class II (HLA-DR). Proposed phenotypic characteristics of MDSCs in several cancers are summarized in Table 4 and Figure 4. Molecules involved in MDSCs mediated suppression of immune cells are summarized in Table 5.

Table 4. Phenotypisation of myeloid-derived suppressor cells in solid and hematological malignancies

Cancer types	Phenotype	References
Renal cell	CD11b+/CD14–/CD15+ (Zea et al)	[89]
non–small-cell lung cancer	CD11b+/CD14–/CD33+/CD15+ (Liu et al), CD11b+/CD33+ (Srivastava et al)	[90, 91]
Head and neck squamous cell carcinoma	CD11b+/CD14–/CD33+ (Corzo et al)	[92]
Breast	CD3-CD14-CD19-HLA-DR- (Almand et al)	[87]
Melanoma	CD3-CD14-CD19-CD56-HLA-DR-CD33+ (Daud et al) CD14+HLA-DR-/low (Poschke et al)	[93, 94]
Hepatocellular carcinoma	CD14+/HLA-DR-/low (Hoechst et al)	[95]
Non-Hodgkin lymphoma	CD14+/HLA-DR-/low (Lin et al)	[96]
Multiple myeloma	CD14+/HLA-DR- (Brimnes et al), CD11b+/CD33+/HLA-DRlow/-/CD33+ (Gorgun et al)	[97, 98]
Myelodysplastic syndrome	CD33+HLA-DR-	[99]

Table 5. Molecules associated with suppressive function of myeloid-derived suppressor cells

Molecules	Functions	References
Arginase and iNOS (inducible nitric oxide synthase)	These two enzymes are released by MDSCs. Arginase depletes the non-essential amino acid L-arginine which leads to inhibition of T cell proliferation. iNOS mediates nitric oxide (NO) production. NO suppresses T cell function via induction of apoptosis, inhibition of MHC II expression and inhibition of STAT-5 and JAK3 function in T cells.	[85], [100]
Reactive oxygen species (ROS)	This is also an important factor of MDSCs that contributes to suppressive activity. ROS release has been documented in tumor bearing mice and cancer patients. Several tumor derived factors including TGF-β, IL-6, IL-3, IL-10 and GM-CSF induce ROS synthesis by MDSCs.	[85]
Peroxynitrite	This molecule also induces MDSCs mediated suppression of T cell function. Accumulation of peroxynitrite is noticed where recruitment of MDSCs occurs at the site of inflammation or immunological reactions. In addition, increased peroxynitrite levels are associated with tumor progression in several cancers. Interaction of peroxynitrite producing MDSCs and T cells leads to nitration of T cell receptors, which leads to T cell unresponsiveness.	[85], [101]
Induction of Treg cells	*In vivo* studies showed that MDSCs can induce *de novo* generation of Treg cells. Induction of Treg cells by MDSCs requires tumor specific T cells together with IFN-γ and IL-10, but independent of NO production. In murine tumor model MDSCs induce generation of Treg cells by CTLA-4 (ovarian tumor) and arginase 1 (lymphoma) molecules. However in contrast, other studies report no association of Treg cell generation with MDSCs.	[102], [103], [104]

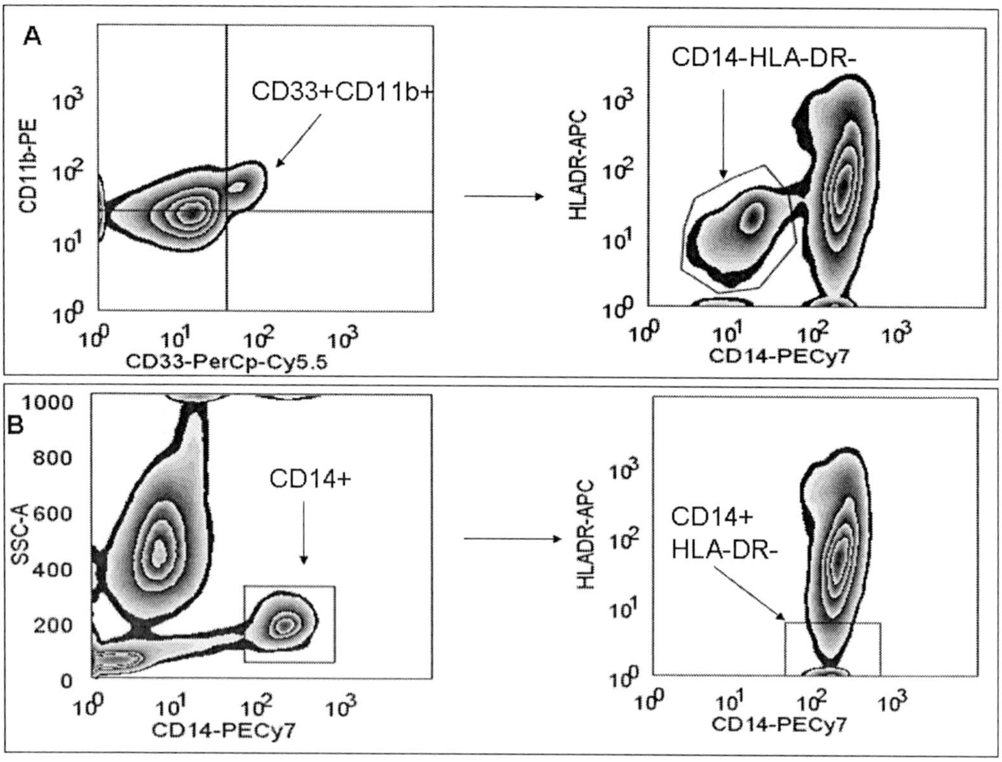

Figure 4. Phenotypisation of granulocytic and monocytic myeloid-derived suppressor cells. As shown in panel (A), granulocytic MDSCs are identified by expression of CD33 and CD11b, and then these positive (CD33+CD11b+) cells are more selectively identified for a negative expression of CD14 and HLA-DR. (B) Monocytic MDSCs are characterized simply based on the expression of CD14 and negative expression of HLA-DR from myeloid cells (CD14+HLA-DR-).

Immune Regulatory Biology of Multiple Myeloma

MM is a malignancy of plasma cells (PCs) that accumulate in BM; thus, leading to several adverse features, such as anemia, hypercalcemia, renal impairment and bone lytic disorders [105]. MM arises from post-germinal centre B cells. These post-germinal centre B cells are activated by T helper cells to differentiate into PCs and to undergo isotype switching and affinity maturation. During this process abnormal DNA recombination events translocate IgH and IgL gene segments into other chromosomes; therefore, abnormal phenotype of PC arises [106]. Apart from adverse features, myeloma patient's BM homeostasis is affected due to infiltration of PCS; thereby, influencing the production of immune cells. Therefore, MM patients are frequently associated with immune dysfunctions and predisposed to various infectious complications [107]. In MM, several studies showed numerical and functional defects of various immune cells, including B cells [108], CD4 and CD8 T cells [109], DCs [110] and NK cells [111]. Also, abnormally reduced numbers of B and T cells strongly associated with poor outcome in MM patients [112, 113]. Myeloma patient's overall survival and disease free survival are severely affected by BM microenvironment that propagates tumor cells; as well by above mentioned abnormalities and impairments. However, introduction of novel drugs (IMiDs and proteosome inhibitors) very substantially improved survival and delayed progression in myeloma patients compared to conventional treatment [114]. Benefits by these novel drugs are not completely translated into better disease free survival and overall survival due to cumulative immune suppression and infections [115].

Relation between immune and cancer systems could be explained by a simple concept called as cancer immunoediting. This concept explains how the immune system is evaded by cancer system. Immunoediting has three phases including elimination phase, equilibrium phase and evasion phase [116, 117]. Elimination phase is highly similar to immune surveillance, where the immune system detects tumor cells and kills them. In the equilibrium phase, tumor cells become dormant, and the immune system selectively destroys susceptible tumor clones and prevents tumor progression. In the escape phase, immune responses fail to suppress the tumor growth which leads to development of tumor variants that are resistant to anti-tumor responses [116, 117].

There are several factors have been proposed that either help the tumor to progress or tumor to evade. In this section, we will try to convince that Treg cells can be a factor that helps in progression of disease through suppressing effector immune cells in MM patients. Also, other studies have shown that a number of molecules were involved (CCL17, CCL22, PGE2) in expansion and recruitment of Treg cells at tumor site or in periphery [118, 119]. We believe that Treg cells dependent immune evasion and tumor dependent expansion of Treg cells could be a vicious cycle in cancer patients that enhances evasion of tumor cells.

In MM, we and others showed that CD4 Treg cells were significantly increased and potentially suppressed effector CD4 T cells. Suppression of effector T cells was noticed in a concentration dependent fashion. Our cytokine profile data showed reduced level of pro-inflammatory cytokine from effector T cells in the presence of CD4 Treg cells. This result further acknowledges suppressive function of CD4 Treg cells [68]. However, two studies have demonstrated that CD4 Treg cells were reduced in myeloma patients. These contrasting data are due to difference in CD4 Treg cell phenotypisation (Table 6). We have also shown that naïve and activated CD4 Treg cells from MM patients are increased and suppressive [68], and recently, another study reported that increased numbers of naïve and memory Treg cells in MM were due to peripherally expanded Treg cells rather than thymus derived Treg cells using TCR excision circle analysis [120]. Beyer at al, have confirmed that expression of molecules (CTLA-4, GITR and CD62L) associated with the function of Treg cells was increased in MM patients compared to healthy donors [120] Previously, a large number of studies have reported prognostic significance of CD4 Treg cells in various cancers, including Hodgkin's lymphoma, follicular lymphoma, breast cancer, gastric malignancies, and ovarian cancers [121, 122, 123, 124, 125]. For the first time in MM, we have shown that increased numbers of Treg cells predicted shorter time to progression (TTP) [68]. Very recently, a study also showed that high number of Treg cells from MM patients was associated with poor survival [126]. However, several questions are yet to be answered in MM immune regulatory biology; do tumor cells enhance the expansion of Treg cells during progression or is progression induced by Treg cells due to suppression of anti-tumor responses? We believe that both processes are mutually inclusive because several *in vitro* and animal model studies in other cancers documented that tumor and tumor infiltrating cells derived factors, such as CCL2, PGE2, H-ferritin, IDO and IL-10 recruit and enhance Treg cells in the tumor microenvironment and thereby, increase the risk of tumor progression [121, 127, 128]. In MM, a study showed that DCs enhanced the expansion of Treg cells *in vitro* and *in vivo*

[129]. Above raised questions were recently addressed by a British study, which clearly showed that myeloma cells induce the CD4 Treg cells in a concentration dependent fashion. These tumor derived Treg cells were polyclonal and found to be suppressive [130].

Apart from prognostic value of CD4 Treg cells, several studies and our data have shown that increased numbers of CD4 Treg cells were significantly associated with adverse features of myeloma such as ISS (international staging system) staging, paraprotein level, hypercalcemia, IgA myeloma subtype and lower normal PC count [68, 131]. These data suggest that increased numbers of suppressive CD4 Treg cells play a significant role in myeloma disease progression and immune impairment. Table 6 summarizes a timeline of research studies focused on Treg cells of MM patients.

We were the first to report the presence of increased numbers of naturally occurring CD8 Treg cells in myeloma patients [78]. Previously, their presence has been reported in solid tumors including prostate, colorectal and nasopharyngeal cancers [77, 79, 80]. More interestingly, we demonstrated that CD8 Treg cells share similar phenotypic features of CD4 Treg cells. These regulatory cells expressed several antigens ranging from dim to high intensity, such as CD28, CD62L, CTLA-4 and CD45RO [78]. As a typical marker of regulatory cells, CD8 Treg cells did not express CD127. This observation was consistent with a previous finding [77]. CTLA-4 expression is always associated with high intensity of Treg cell mediated suppression [33]. Our study showed that CTLA-4 expression in MM patients was significantly increased compared to healthy donors. Also, expression of CD45RO and CD62L demonstrates activated phenotype of CD8 Treg cells [72, 132]. This finding suggests that CD8 Treg cells might be a very active suppressive subset in MM patients. We used CFSE based proliferation assay to assess the suppressive function of CD8 Treg cells and validated their suppressive function with cytokine profile (IFN-γ and IL-10) [78]. Proliferation assay findings clearly showed in line with other findings (prostate, nasopharyngeal and colorectal cancer) that CD8 Treg cells inhibited proliferation of CD4 T cells in a concentration dependent manner [77, 79, 133]. In the absence of CD8 Treg cells, CD4 T cells isolated from MM patients and healthy donors proliferated similarly in proliferation assays [78]. These observations suggest no defects in proliferation potential of CD4 T cells from MM patients but immune suppression in MM patients can be induced in the presence of CD8 Treg cells. In line with the finding of inhibition of proliferation by CD8 Treg cells, we and others have confirmed that pro-inflammatory cytokine (IFN-γ) level was decreased with respect to the number of CD8 Treg cells added to the proliferation assay [77, 78]. We believe that from our and Chaput et al. studies, pro-inflammatory cytokine response by T cells could be reduced due to the presence of CD8 Treg cells [77]. In addition, to prove the immunosuppressive activity of CD8 Treg cells, IL-10 cytokine was evaluated from culture supernatants. The data showed that depending on the number of CD8 Treg cells added to proliferation assay, level of IL-10 was increased in MM patients and healthy donors. Data from proliferation assays without the presence of CD8 Treg cells showed significant increase in the level of IL-10 from MM patients compared to healthy donors. However, proliferation experiments showed no defect in proliferation potential of CD4 T cells from MM patients in the absence of CD8 Treg cells but we believe that these cells are polarized toward the Th2 cell type, which is evidenced by increased level of IL-10 compared to healthy donors. This observation further supports the idea of immunological impairment in MM patients [78]. Immune impairment induced by CD8 Treg cells was also evidenced by correlation analysis in MM patients. CD4/CD8 ratio and total lymphocyte counts were negatively associated with CD8 Treg cell numbers [78]. Clinical significance of CD8 Treg cells in MM patients are yet to be revealed.

Very recently, few groups have been focused to study the role of MDSCs in MM. First data of MDSCs in myeloma were published by a Danish group [97]. They have studied only numerical changes of monocytic MDSCs in MM and showed that myeloma patients had increased numbers of monocytic MDSCs than healthy donors. Later, a most convincing data were published from Dana Farber Cancer Institute [98]. Gorgun et al., have shown that numbers of granulocytic MDSCs in MM patients were increased compared to healthy donors [98]. *In vitro* data from Gorgun et al., suggested that MDSCs initiated myeloma cell growth and suppressed immune cells significantly [98]. It was also confirmed that interaction between MDSCs and myeloma cells is bidirectional. Novel drugs such as lenalidomide and bortezomib were not able to overcome MDSCs induced immune suppression [98]. We have also observed similar results *in vivo*, patients treated with lenalidomide plus dexamethasone had significantly increased numbers of suppressive MDSCs compared to baseline numbers [134]. Our data also showed clinical significance of MDSCs with regards to treatment response; patients who achieved minimal response had increased numbers of MDSCs after treatment compared to baseline but no changes were observed in patients who achieved greater than minimal response.

Studying these immune regulatory (CD4 and CD8 Treg cells) and suppressor cells (MDSCs) revealed the complexity of immune suppression in myeloma patients. It seems that tumor progression and immune suppression is a mutual process in myeloma. Therefore, targeting or depletion of immune regulatory and suppressor cells in cancer patients could overcome the immune suppression and immune suppression-mediated disease progression. Based on preclinical models, clinical trials have been started to target Treg cells using denileukin diftitox, a fusion protein of diphtheria toxin and IL-2 in several malignancies, including metastatic renal cell carcinoma (denileukin diftitox plus tumor RNA transfected DCs) and colorectal carcinoma (denileukin diftitox plus DCs expressing carcinoembryonic antigen) [135, 136, 137].

Table 6. T regulatory cells timeline in multiple myeloma

Study and year	Treg cell phenotype and their numbers	Treg cells mechanistic data	Clinical significance
Prabhala et al., 2006 [138]	CD4+FoxP3+, significantly reduced	Lack suppressive function significantly	Not shown
Beyer et al., 2006 [120]	CD4+CD25hi+ or CD4+CD25hi+FoxP3+, significantly increased	Concentration dependent inhibition of responder cells (CD4+CD25-) and their production of IFN-γ. Suppression was similar between healthy donors and myeloma patients	Not shown
Atanackovic et al., 2008 [139]	CD4+CD25hi+FoxP3+, significantly increased in post-allogenic transplanted patients than healthy donors and newly diagnosed patients	Suppressive	Reduced Treg cells were observed in patients with graft versus host disease (GVHD) than patients without GVHD.

Table 6. (Continued)

Study and year	Treg cell phenotype and their numbers	Treg cells mechanistic data	Clinical significance
Feyler et al., 2009 [131]	CD4+CD25hi+FoxP3+, significantly increased	Suppressive	Increased number of Treg cells associated with disease stage.
Brimnes et al., 2010 [97]	CD4+FoxP3+, significantly increased	Suppressive, inhibited proliferation of both CD4 and CD8 T cells	Not shown
Gupta et al., 2011 [140]	CD4+CD25hi+FoxP3+CD127-, significantly reduced	Concentration dependent inhibition of responder cells (CD4+CD25-)	Reduced number of Treg cells associated with disease stage.
Feyler et al., 2012 [130]	CD4+CD25+FoxP3+	Using an *in vitro* model, it has been shown that myeloma cells induce generation of Treg cells from effector T cells (CD4+CD25-) in an ICOS molecule dependent manner but not with IL-10/TGF-β. Tumor generated Treg cells were polyclonal and immunosuppressive	Not shown
Giannopoulos et al., 2012 [126]	CD4+CD25hi+FoxP3+, significantly increased	Not shown	Increased number of Treg cells significantly correlated with worse overall survival compared to patients with lower number of Treg cells
Muthu Raja et al., 2012 [68]	CD4+CD25hi+FoxP3+, significantly increased. As well, naive (CD4+CD45RA+FoxP3dim+) and activated (CD4+CD45RA-FoxP3+) Treg cells were significantly increased	Suppressive, inhibited proliferation of responder cells in a concentration dependent fashion. Naive and activated Treg cells also suppressed proliferation of naive CD4 T cells	Predicted progression risk in myeloma patients
Muthu Raja et al., 2012 [78]	CD8+CD25hi+, significantly increased. Increase in FoxP3 expression was confirmed by flow cytometry and RT-PCR	Suppressive, inhibited proliferation of responder cells and their cytokine secretion in a concentration dependent fashion	Not shown
Guichelaar et al., 2013 [141]	CD4+CD127dim+FoxP3+	In immunodeficient mice models co-infusion of Treg cells did not impair the anti-tumor response in the BM microenvironment. This is due to neutralizing effects of BMSCs via secretion of IL-6/IL-1β and converting to pro-inflammatory cells (Th17). Whereas, Treg cells outside of BM suppressed anti-tumor response	Therapeutic benefit

Data from these trials showed improved immune stimulation, reduced numbers of Treg cells and cytotoxic T cell responses. Recently, a phase II trial in MM has been involved in targeting a PD-1 molecule (associated with Treg cells) using a monoclonal antibody (CT-011) in combination with/without DC vaccination (Trial no. NCT01067287). Data from this trial are yet to be published.

Conclusion

Flow cytometry serves as an indispensable tool in monitoring of immune cells in several pathological conditions. During the past decade, monitoring of Treg cells using flow cytometry in solid and hematological malignancies has become obvious, due to their significance in revealing immune and prognostic status. From our experience, we suggest that monitoring of Treg cells and their subsets in MM patients must be included in routine clinical laboratory analysis due to their prognostic significance and their association with adverse clinical features. Since MM is a complex disease, immune system is highly deregulated in these patients. A large number of studies in MM showed expansion of several immune suppressor cells, including Treg cells and MDSCs. These cells have been proven to play a significant role in suppression of effector immune response and promotion of tumor growth; thereby, disease progression and infectious complications occur. Apart from routine monitoring of Treg cells, flow cytometry serves as a precise tool in functional characterization of Treg cells compared to thymidine uptake method.

Acknowledgment

Supported by grants: The Ministry of Education, Youth and Sports; MSM0021622434, Grant Agency of Czech Science Foundation; GAP304/10/1395 and The Ministry of Health, Czech Republic - conceptual development of research organization; FNBr, 65269705.

References

[1] Sakaguchi S, Sakaguchi N, Asano M, Itoh M, Toda M (1995) Immunologic self-tolerance maintained by activated T cells expressing IL-2 receptor alpha-chains (CD25). Breakdown of a single mechanism of self-tolerance causes various autoimmune diseases. *J. Immunol.* 155: 1151-1164.

[2] Sakaguchi S, Takahashi T, Nishizuka Y (1982) Study on cellular events in post-thymectomy autoimmune oophoritis in mice. II. Requirement of Lyt-1 cells in normal female mice for the prevention of oophoritis. *J. Exp. Med.* 156: 1577-1586.

[3] North R. J (1985) Down-regulation of the antitumor immune response. *Advances in cancer research* 45: 1-43.

[4] Chakraborty N. G, Twardzik D. R, Sivanandham M, Ergin M, Hellstrom K. E, et al. (1990) Autologous melanoma-induced activation of regulatory T cells that suppress cytotoxic response. *The Journal of Immunology* 145: 2359-2364.

[5] Chattopadhyay S, Chakraborty N. G, Mukherji B (2005) Regulatory T cells and tumor immunity. *Cancer Immunology Immunotherapy* 54: 1153-1161.

[6] Sakaguchi S, Wing K, Miyara M (2007) Regulatory T cells - a brief history and perspective. *European journal of immunology* 37 Suppl 1: S116-123.

[7] Itoh M, Takahashi T, Sakaguchi N, Kuniyasu Y, Shimizu J, et al. (1999) Thymus and autoimmunity: Production of CD25(+)CD4(+) naturally anergic and suppressive T cells as a key function of the thymus in maintaining immunologic self-tolerance. *Journal of Immunology* 162: 5317-5326.

[8] Lahl K, Loddenkemper C, Drouin C, Freyer J, Arnason J, et al. (2007) Selective depletion of Foxp3(+) regulatory T cells induces a scurfy-like disease. *Journal of Experimental Medicine* 204: 57-63.

[9] Fontenot J. D, Gavin M. A, Rudensky A. Y (2003) Foxp3 programs the development and function of CD4+CD25+ regulatory T cells. *Nat. Immunol.* 4: 330-336.

[10] Tuovinen H, Pekkarinen P. T, Rossi L. H, Mattila I, Arstila T. P (2008) The FOXP3(+) subset of human CD4(+)CD8(+) thymocytes is immature and subject to intrathymic selection. *Immunology and Cell Biology* 86: 523-529.

[11] Lee H. M, Hsieh C. S (2009) Rare Development of Foxp3(+) Thymocytes in the CD4(+)CD8(+) Subset. *Journal of Immunology* 183: 2261-2266.

[12] Sakaguchi S, Yamaguchi T, Nomura T, Ono M (2008) Regulatory T cells and immune tolerance. *Cell* 133: 775-787.

[13] Picca C. C, Larkin J, Boesteanu A, Lerman M. A, Rankin A. L, et al. (2006) Role of TCR specificity in CD4(+)CD25(+) regulatory T-cell selection. *Immunological Reviews* 212: 74-85.

[14] Malek T. R, Yu A. X, Zhu L. J, Matsutani T, Adeegbe D, et al. (2008) IL-2 Family of Cytokines in T Regulatory Cell Development and Homeostasis. *Journal of Clinical Immunology* 28: 635-639.

[15] Watanabe N, Hanabuchi S, Soumelis V, Yuan W, Ho S, et al. (2004) Human thymic stromal lymphopoietin promotes dendritic cell-mediated CD4+ T cell homeostatic expansion. *Nature immunology* 5: 426-434.

[16] Hanabuchi S, Lto T, Park W. R, Watanabe N, Shaw J. L, et al. (2010) Thymic Stromal Lymphopoietin-Activated Plasmacytoid Dendritic Cells Induce the Generation of FOXP3(+) Regulatory T Cells in Human Thymus. *Journal of Immunology* 184: 2999-3007.

[17] Watanabe N, Wang Y. H, Lee H. K, Ito T, Cao W, et al. (2005) Hassall's corpuscles instruct dendritic cells to induce CD4+CD25+ regulatory T cells in human thymus. *Nature* 436: 1181-1185.

[18] Liu Y. J, Soumelis V, Watanabe N, Ito T, Wang Y. H, et al. (2007) TSLP: An epithelial cell cytokine that regulates T cell differentiation by conditioning dendritic cell maturation. *Annual Review of Immunology* 25: 193-219.

[19] Stites D. P, Pavia CS (1979) Ontogeny of Human T-Cells. *Pediatrics* 64: 795-802.

[20] Darrasse-Jeze G, Marodon G, Salomon B. L, Catala M, Klatzmann D (2005) Ontogeny of CD4(+)CD25(+) regulatory/suppressor T cells in human fetuses. *Blood* 105: 4715-4721.

[21] Bennett C. L, Christie J, Ramsdell F, Brunkow M. E, Ferguson P. J, et al. (2001) The immune dysregulation, polyendocrinopathy, enteropathy, X-linked syndrome (IPEX) is caused by mutations of FOXP3. *Nature Genetics* 27: 20-21.

[22] Brunkow M. E, Jeffery E. W, Hjerrild K. A, Paeper B, Clark L. B, et al. (2001) Disruption of a new forkhead/winged-helix protein, scurfin, results in the fatal lymphoproliferative disorder of the scurfy mouse. *Nature Genetics* 27: 68-73.

[23] Chatila T. A, Blaeser F, Ho N, Lederman H. M, Voulgaropoulos C, et al. (2000) JM2, encoding a fork head-related protein, is mutated in X-linked autoimmunity-allergic disregulation syndrome. *Journal of Clinical Investigation* 106: R75-R81.

[24] Hori S, Nomura T, Sakaguchi S (2003) Control of regulatory T cell development by the transcription factor Foxp3. *Science* 299: 1057-1061.

[25] Allan S. E, Passerini L, Bacchetta R, Crellin N, Dai M, et al. (2005) The role of 2 FOXP3 isoforms in the generation of human CD4+ Tregs. *J. Clin. Invest.* 115: 3276-3284.

[26] Yagi H, Nomura T, Nakamura K, Yamazaki S, Kitawaki T, et al. (2004) Crucial role of FOXP3 in the development and function of human CD25(+)CD4(+) regulatory T cells. *International Immunology* 16: 1643-1656.

[27] Lopes J. E, Torgerson T. R, Schubert L. A, Anover S. D, Ocheltree E. L, et al. (2006) Analysis of FOXP3 reveals multiple domains required for its function as a transcriptional repressor. *Journal of Immunology* 177: 3133-3142.

[28] Bettelli E, Dastrange M, Oukka M (2005) Foxp3 interacts with nuclear factor of activated T cells and NF-kappa B to repress cytokine gene expression and effector functions of T helper cells. *Proceedings of the National Academy of Sciences of the United States of America* 102: 5138-5143.

[29] Walker M. R, Kasprowicz D. J, Gersuk V. H, Benard A, Van Landeghen M, et al. (2003) Induction of FoxP3 and acquisition of T regulatory activity by stimulated human CD4(+)CD25(-) T cells. *Journal of Clinical Investigation* 112: 1437-1443.

[30] Hoffmann P, Boeld T. J, Eder R, Huehn J, Floess S, et al. (2009) Loss of FOXP3 expression in natural human CD4(+)CD25(+) regulatory T cells upon repetitive in vitro stimulation. *European journal of immunology* 39: 1088-1097.

[31] Allan S. E, Song-Zhao G. X, Abraham T, McMurchy A. N, Levings M. K (2008) Inducible reprogramming of human T cells into Treg cells by a conditionally active form of FOXP3. *European journal of immunology* 38: 3282-3289.

[32] Tran D. Q, Ramsey H, Shevach E. M (2007) Induction of FOXP3 expression in naive human CD4+FOXP3 T cells by T-cell receptor stimulation is transforming growth factor-beta dependent but does not confer a regulatory phenotype. *Blood* 110: 2983-2990.

[33] Sakaguchi S, Miyara M, Costantino C. M, Hafler D. A (2010) FOXP3+ regulatory T cells in the human immune system. *Nature reviews Immunology* 10: 490-500.

[34] Mempel T. R, Pittet M. J, Khazaie K, Weninger W, Weissleder R, et al. (2006) Regulatory T cells reversibly suppress cytotoxic T cell function independent of effector differentiation. *Immunity* 25: 129-141.

[35] Piccirillo C. A, Shevach E. M (2001) Cutting edge: Control of CD8(+) T cell activation by CD4(+)CD25(+) immunoregulatory cells. *Journal of Immunology* 167: 1137-1140.

[36] Bopp T, Becker C, Klein M, Klein-Hessling S, Palmetshofer A, et al. (2007) Cyclic adenosine monophosphate is a key component of regulatory T cell-mediated suppression. *J. Exp. Med.* 204: 1303-1310.

[37] Grossman W. J, Verbsky J. W, Barchet W, Colonna M, Atkinson J. P, et al. (2004) Human T regulatory cells can use the perforin pathway to cause autologous target cell death. *Immunity* 21: 589-601.

[38] Gondek D. C, Lu L. F, Quezada S. A, Sakaguchi S, Noelle R. J (2005) Cutting edge: contact-mediated suppression by CD4+CD25+ regulatory cells involves a granzyme B-dependent, perforin-independent mechanism. *J. Immunol.* 174: 1783-1786.

[39] Zhao D. M, Thornton A. M, DiPaolo R. J, Shevach E. M (2006) Activated CD4(+)CD25(+) T cells selectively kill B lymphocytes. *Blood* 107: 3925-3932.

[40] Tang Q, Bluestone J. A (2008) The Foxp3+ regulatory T cell: a jack of all trades, master of regulation. *Nat. Immunol.* 9: 239-244.

[41] Mellor A. L, Munn D. H (2004) IDO expression by dendritic cells: tolerance and tryptophan catabolism. *Nat. Rev. Immunol.* 4: 762-774.

[42] Munn D. H, Mellor A. L (2004) IDO and tolerance to tumors. *Trends in Molecular Medicine* 10: 15-18.

[43] Puccetti P, Grohmann U (2007) IDO and regulatory T cells: a role for reverse signalling and non-canonical NF-kappa B activation. *Nature Reviews Immunology* 7: 817-823.

[44] Tang Q, Boden E. K, Henriksen K. J, Bour-Jordan H, Bi M, et al. (2004) Distinct roles of CTLA-4 and TGF-beta in CD4+CD25+ regulatory T cell function. *European journal of immunology* 34: 2996-3005.

[45] Thornton A. M, Shevach E. M (1998) CD4+CD25+ immunoregulatory T cells suppress polyclonal T cell activation in vitro by inhibiting interleukin 2 production. *The Journal of experimental medicine* 188: 287-296.

[46] Read S, Greenwald R, Izcue A, Robinson N, Mandelbrot D, et al. (2006) Blockade of CTLA-4 on CD4(+) CD25(+) regulatory T cells abrogates their function in vivo. *Journal of Immunology* 177: 4376-4383.

[47] Chikuma S, Bluestone J. A (2007) Expression of CTLA-4 and FOXP3 in cis protects from lethal lymphoproliferative disease. *European journal of immunology* 37: 1285-1289.

[48] Chen W, Jin W, Hardegen N, Lei K. J, Li L, et al. (2003) Conversion of peripheral CD4+CD25- naive T cells to CD4+CD25+ regulatory T cells by TGF-beta induction of transcription factor Foxp3. *J. Exp. Med.* 198: 1875-1886.

[49] Ito T, Hanabuchi S, Wang Y. H, Park W. R, Arima K, et al. (2008) Two functional subsets of FOXP3+ regulatory T cells in human thymus and periphery. *Immunity* 28: 870-880.

[50] Choi B. M, Pae H. O, Jeong Y. R, Kim Y. M, Chung H. T (2005) Critical role of heme oxygenase-1 in Foxp3-mediated immune suppression. *Biochemical and Biophysical Research Communications* 327: 1066-1071.

[51] Xia Z. W, Zhong W. W, Xu L. Q, Sun J. L, Shen Q. X, et al. (2006) Heme oxygenase-1-mediated CD4(+)CD25(high) regulatory T cells suppress allergic airway inflammation. *Journal of Immunology* 177: 5936-5945.

[52] Collison L. W, Workman C. J, Kuo T. T, Boyd K, Wang Y, et al. (2007) The inhibitory cytokine IL-35 contributes to regulatory T-cell function. *Nature* 450: 566-569.

[53] Baecher-Allan C, Brown J. A, Freeman G. J, Hafler D. A (2001) CD4+CD25high regulatory cells in human peripheral blood. *J. Immunol.* 167: 1245-1253.

[54] Allan S. E, Crome S. Q, Crellin N. K, Passerini L, Steiner T. S, et al. (2007) Activation-induced FOXP3 in human T effector cells does not suppress proliferation or cytokine production. *International Immunology* 19: 345-354.

[55] Liu W, Putnam A. L, Xu-Yu Z, Szot G. L, Lee M. R, et al. (2006) CD127 expression inversely correlates with FoxP3 and suppressive function of human CD4+ T reg cells. *J. Exp. Med.* 203: 1701-1711.

[56] Seddiki N, Santner-Nanan B, Martinson J, Zaunders J, Sasson S, et al. (2006) Expression of interleukin (IL)-2 and IL-7 receptors discriminates between human regulatory and activated T cells. *J. Exp. Med.* 203: 1693-1700.

[57] Mazzucchelli R, Durum S. K (2007) Interleukin-7 receptor expression: intelligent design. *Nature Reviews Immunology* 7: 144-154.

[58] Aerts N. E, Dombrecht E. J, Ebo D. G, Bridts C. H, Stevens W. J, et al. (2008) Activated T cells complicate the identification of regulatory T cells in rheumatoid arthritis. *Cellular Immunology* 251: 109-115.

[59] Hamann A, Klugewitz K, Austrup F, Jablonski-Westrich D (2000) Activation induces rapid and profound alterations in the trafficking of T cells. *European journal of immunology* 30: 3207-3218.

[60] Miyara M, Yoshioka Y, Kitoh A, Shima T, Wing K, et al. (2009) Functional delineation and differentiation dynamics of human CD4+ T cells expressing the FoxP3 transcription factor. *Immunity* 30: 899-911.

[61] Curotto de Lafaille M. A, Lafaille J. J (2009) Natural and adaptive foxp3+ regulatory T cells: more of the same or a division of labor? *Immunity* 30: 626-635.

[62] Gambineri E, Torgerson T. R, Ochs H. D (2003) Immune dysregulation, polyendocrinopathy, enteropathy, and X-linked inheritance (IPEX), a syndrome of systemic autoimmunity caused by mutations of FOXP3, a critical regulator of T-cell homeostasis. *Curr. Opin. Rheumatol.* 15: 430-435.

[63] Groux H, O'Garra A, Bigler M, Rouleau M, Antonenko S, et al. (1997) A CD4+ T-cell subset inhibits antigen-specific T-cell responses and prevents colitis. *Nature* 389: 737-742.

[64] Vieira P. L, Christensen J. R, Minaee S, O'Neill E. J, Barrat F. J, et al. (2004) IL-10-secreting regulatory T cells do not express Foxp3 but have comparable regulatory function to naturally occurring CD4+CD25+ regulatory T cells. *J. Immunol.* 172: 5986-5993.

[65] Apostolou I, von Boehmer H (2004) In vivo instruction of suppressor commitment in naive T cells. *J. Exp. Med.* 199: 1401-1408.

[66] Fischer K, Voelkl S, Heymann J, Przybylski G. K, Mondal K, et al. (2005) Isolation and characterization of human antigen-specific TCR alpha beta+ CD4(-)CD8- double-negative regulatory T cells. *Blood* 105: 2828-2835.

[67] Peng G, Wang H. Y, Peng W, Kiniwa Y, Seo K. H, et al. (2007) Tumor-infiltrating gammadelta T cells suppress T and dendritic cell function via mechanisms controlled by a unique toll-like receptor signaling pathway. *Immunity* 27: 334-348.

[68] Muthu Raja K. R, Rihova L, Zahradova L, Klincova M, Penka M, et al. (2012) Increased T regulatory cells are associated with adverse clinical features and predict progression in multiple myeloma. *PloS one* 7: e47077.

[69] Kruisbeek A. M, Shevach E, Thornton A. M (2004) Proliferative assays for T cell function. Current protocols in immunology / edited by John E Coligan [et al] Chapter 3: Unit 3 12.

[70] Wang R. F (2008) CD8+ regulatory T cells, their suppressive mechanisms, and regulation in cancer. *Hum. Immunol.* 69: 811-814.

[71] Endharti A. T, Rifa'I M, Shi Z, Fukuoka Y, Nakahara Y, et al. (2005) Cutting edge: CD8+CD122+ regulatory T cells produce IL-10 to suppress IFN-gamma production and proliferation of CD8+ T cells. *J. Immunol.* 175: 7093-7097.

[72] Cosmi L, Liotta F, Lazzeri E, Francalanci M, Angeli R, et al. (2003) Human CD8+CD25+ thymocytes share phenotypic and functional features with CD4+CD25+ regulatory thymocytes. *Blood* 102: 4107-4114.

[73] Tang X. L, Smith T. R, Kumar V (2005) Specific control of immunity by regulatory CD8 T cells. *Cell Mol. Immunol.* 2: 11-19.

[74] Cortesini R, LeMaoult J, Ciubotariu R, Cortesini N. S (2001) CD8+CD28- T suppressor cells and the induction of antigen-specific, antigen-presenting cell-mediated suppression of Th reactivity. *Immunol. Rev.* 182: 201-206.

[75] Suciu-Foca N, Manavalan J. S, Scotto L, Kim-Schulze S, Galluzzo S, et al. (2005) Molecular characterization of allospecific T suppressor and tolerogenic dendritic cells: review. *Int. Immunopharmacol.* 5: 7-11.

[76] Chang C. C, Ciubotariu R, Manavalan J. S, Yuan J, Colovai A. I, et al. (2002) Tolerization of dendritic cells by T(S) cells: the crucial role of inhibitory receptors ILT3 and ILT4. *Nat. Immunol.* 3: 237-243.

[77] Chaput N, Louafi S, Bardier A, Charlotte F, Vaillant J. C, et al. (2009) Identification of CD8+CD25+Foxp3+ suppressive T cells in colorectal cancer tissue. *Gut* 58: 520-529.

[78] Muthu Raja K. R, Kubiczkova L, Rihova L, Piskacek M, Vsianska P, et al. (2012) Functionally Suppressive CD8 T Regulatory Cells Are Increased in Patients with Multiple Myeloma: A Cause for Immune Impairment. *PloS one* 7: e49446.

[79] Kiniwa Y, Miyahara Y, Wang H. Y, Peng W, Peng G, et al. (2007) CD8+ Foxp3+ regulatory T cells mediate immunosuppression in prostate cancer. *Clin. Cancer Res.* 13: 6947-6958.

[80] Li J, Huang Z. F, Xiong G, Mo H. Y, Qiu F, et al. (2011) Distribution, characterization, and induction of CD8(+) regulatory T cells and IL-17-producing CD8(+) T cells in nasopharyngeal carcinoma. *Journal of Translational Medicine* 9.

[81] Cosmi L, Liotta F, Angeli R, Mazzinghi B, Santarlasci V, et al. (2004) Th2 cells are less susceptible than Th1 cells to the suppressive activity of CD25(+) regulatory thymocytes because of their responsiveness to different cytokines. *Blood* 103: 3117-3121.

[82] Xystrakis E, Dejean A. S, Bernard I, Druet P, Liblau R, et al. (2004) Identification of a novel natural regulatory CD8 T-cell subset and analysis of its mechanism of regulation. *Blood* 104: 3294-3301.

[83] Sarantopoulos S, Lu L, Cantor H (2004) Qa-1 restriction of CD8+ suppressor T cells. *J. Clin. Invest.* 114: 1218-1221.

[84] Kim H-J, Verbinnen B, Tang X, Lu L, Cantor H (2010) Inhibition of follicular T-helper cells by CD8+ regulatory T cells is essential for self tolerance. *Nature* 467: 328-332.

[85] Gabrilovich D. I, Nagaraj S (2009) Myeloid-derived suppressor cells as regulators of the immune system. *Nat. Rev. Immunol.* 9: 162-174.

[86] Youn J. I, Nagaraj S, Collazo M, Gabrilovich D. I (2008) Subsets of myeloid-derived suppressor cells in tumor-bearing mice. *J. Immunol.* 181: 5791-5802.

[87] Almand B, Clark J. I, Nikitina E, van Beynen J, English N. R, et al. (2001) Increased production of immature myeloid cells in cancer patients: a mechanism of immunosuppression in cancer. *J. Immunol.* 166: 678-689.

[88] Diaz-Montero C. M, Salem M. L, Nishimura M. I, Garrett-Mayer E, Cole D. J, et al. (2009) Increased circulating myeloid-derived suppressor cells correlate with clinical cancer stage, metastatic tumor burden, and doxorubicin-cyclophosphamide chemotherapy. *Cancer Immunol. Immunother.* 58: 49-59.

[89] Zea A. H, Rodriguez P. C, Atkins N. B, Hernandez C, Signoretti S, et al. (2005) Arginase-producing myeloid suppressor cells in renal cell carcinoma patients: A mechanism of tumor evasion. *Cancer Research* 65: 3044-3048.

[90] Liu C. Y, Wang Y. M, Wang C. L, Feng P. H, Ko H. W, et al. (2010) Population alterations of L-arginase- and inducible nitric oxide synthase-expressed CD11b+/CD14(-)/CD15+/CD33+ myeloid-derived suppressor cells and CD8+ T lymphocytes in patients with advanced-stage non-small cell lung cancer. *Journal of cancer research and clinical oncology* 136: 35-45.

[91] Srivastava M. K, Bosch J. J, Thompson J. A, Ksander B. R, Edelman M. J, et al. (2008) Lung cancer patients' CD4(+) T cells are activated in vitro by MHC II cell-based vaccines despite the presence of myeloid-derived suppressor cells. *Cancer immunology, immunotherapy : CII* 57: 1493-1504.

[92] Corzo C. A, Cotter M. J, Cheng P, Cheng F, Kusmartsev S, et al. (2009) Mechanism regulating reactive oxygen species in tumor-induced myeloid-derived suppressor cells. *Journal of immunology* 182: 5693-5701.

[93] Daud A. I, Mirza N, Lenox B, Andrews S, Urbas P, et al. (2008) Phenotypic and functional analysis of dendritic cells and clinical outcome in patients with high-risk melanoma treated with adjuvant granulocyte macrophage colony-stimulating factor. *Journal of clinical oncology : official journal of the American Society of Clinical Oncology* 26: 3235-3241.

[94] Poschke I, Mougiakakos D, Hansson J, Masucci G. V, Kiessling R (2010) Immature immunosuppressive CD14+HLA-DR-/low cells in melanoma patients are Stat3hi and overexpress CD80, CD83, and DC-sign. *Cancer research* 70: 4335-4345.

[95] Hoechst B, Ormandy L. A, Ballmaier M, Lehner F, Kruger C, et al. (2008) A new population of myeloid-derived suppressor cells in hepatocellular carcinoma patients induces CD4(+)CD25(+)Foxp3(+) T cells. *Gastroenterology* 135: 234-243.

[96] Lin Y, Gustafson M. P, Bulur P. A, Gastineau D. A, Witzig T. E, et al. (2011) Immunosuppressive CD14+HLA-DR(low)/- monocytes in B-cell non-Hodgkin lymphoma. *Blood* 117: 872-881.

[97] Brimnes M. K, Vangsted A. J, Knudsen L. M, Gimsing P, Gang A. O, et al. (2010) Increased level of both CD4+FOXP3+ regulatory T cells and CD14+HLA-DR$^-$/low myeloid-derived suppressor cells and decreased level of dendritic cells in patients with multiple myeloma. *Scand. J. Immunol.* 72: 540-547.

[98] Gorgun G. T, Whitehill G, Anderson J. L, Hideshima T, Maguire C, et al. (2013) Tumor-promoting immune-suppressive myeloid-derived suppressor cells in the multiple myeloma microenvironment in humans. *Blood* 121: 2975-2987.

[99] Wei S, Chen X. H, Rocha K, Qi D. H, Tao J. G, et al. (2009) Myeloid-Derived Suppressor Cells (MDSC) Are Effectors of Bone Marrow Suppression in Lower Risk Myelodysplastic Syndromes (MDS). *Blood* 114: 247-248.

[100] Rodriguez P. C, Zea A. H, Culotta K. S, Zabaleta J, Ochoa J. B, et al. (2002) Regulation of T cell receptor CD3zeta chain expression by L-arginine. *J. Biol. Chem.* 277: 21123-21129.

[101] Nagaraj S, Gupta K, Pisarev V, Kinarsky L, Sherman S, et al. (2007) Altered recognition of antigen is a mechanism of CD8+ T cell tolerance in cancer. *Nat. Med.* 13: 828-835.

[102] Yang R, Cai Z, Zhang Y, Yutzy W. H, Roby K. F, et al. (2006) CD80 in immune suppression by mouse ovarian carcinoma-associated Gr-1+CD11b+ myeloid cells. *Cancer Res.* 66: 6807-6815.

[103] Huang B, Pan P. Y, Li Q, Sato A. I, Levy D. E, et al. (2006) Gr-1+CD115+ immature myeloid suppressor cells mediate the development of tumor-induced T regulatory cells and T-cell anergy in tumor-bearing host. *Cancer Res.* 66: 1123-1131.

[104] Serafini P, Mgebroff S, Noonan K, Borrello I (2008) Myeloid-derived suppressor cells promote cross-tolerance in B-cell lymphoma by expanding regulatory T cells. *Cancer Res.* 68: 5439-5449.

[105] Kyle R. A, Rajkumar S. V (2008) Multiple myeloma. *Blood* 111: 2962-2972.

[106] Kuehl W. M, Bergsagel P. L (2002) Multiple myeloma: evolving genctic events and host interactions. *Nature Reviews Cancer* 2: 175-187.

[107] Pratt G, Goodyear O, Moss P (2007) Immunodeficiency and immunotherapy in multiple myeloma. *Br. J. Haematol.* 138: 563-579.

[108] Rawstron A. C, Davies F. E, Owen R. G, English A, Pratt G, et al. (2001) B - lymphocyte suppression in multiple myeloma is a reversible phenomenon specific to normal B - cell progenitors and plasma cell precursors. *British Journal of Haematology* 100: 176-183.

[109] Maecker B, Anderson K. S, von Bergwelt-Baildon M. S, Weller E, Vonderheide R. H, et al. (2003) Viral antigen-specific CD8(+) T-cell responses are impaired in multiple myeloma. *British Journal of Haematology* 121: 842-848.

[110] Brown R. D, Pope B, Murray A, Esdale W, Sze D. M, et al. (2001) Dendritic cells from patients with myeloma are numerically normal but functionally defective as they fail to up-regulate CD80 (B7-1) expression after huCD40LT stimulation because of inhibition by transforming growth factor-beta(1) and interleukin-10. *Blood* 98: 2992-2998.

[111] Carbone E, Neri P, Mesuraca M, Fulciniti M. T, Otsuki T, et al. (2005) HLA class I, NKG2D, and natural cytotoxicity receptors regulate multiple myeloma cell recognition by natural killer cells. *Blood* 105: 251-258.

[112] Kay N. E, Leong T. L, Bone N, Vesole D. H, Greipp P. R, et al. (2001) Blood levels of immune cells predict survival in myeloma patients: results of an Eastern Cooperative Oncology Group phase 3 trial for newly diagnosed multiple myeloma patients. *Blood* 98: 23-28.

[113] Kay N. E, Leong T, Kyle R. A, Greipp P, Billadeau D, et al. (1997) Circulating blood B cells in multiple myeloma: Analysis and relationship to circulating clonal cells and clinical parameters in a cohort of patients entered on the Eastern Cooperative Oncology Group phase III E9486 clinical trial. *Blood* 90: 340-345.

[114] Palumbo A, Rajkumar S. V (2009) Treatment of newly diagnosed myeloma. *Leukemia* 23: 449-456.

[115] Nucci M, Anaissie E (2009) Infections in patients with multiple myeloma in the era of high-dose therapy and novel agents. *Clinical infectious diseases : an official publication of the Infectious Diseases Society of America* 49: 1211-1225.

[116] Raja K. R. M, Hajek R (2011) Regulatory Cells and Multiple Myeloma. MULTIPLE MYELOMA–AN OVERVIEW: 33.

[117] Swann J. B, Smyth M. J (2007) Immune surveillance of tumors. *J. Clin. Invest.* 117: 1137-1146.

[118] Ishida T, Ishii T, Inagaki A, Yano H, Komatsu H, et al. (2006) Specific recruitment of CC chemokine receptor 4-positive regulatory T cells in Hodgkin lymphoma fosters immune privilege. *Cancer Res.* 66: 5716-5722.

[119] Yang Z. Z, Novak A. J, Stenson M. J, Witzig T. E, Ansell S. M (2006) Intratumoral CD4+CD25+ regulatory T-cell-mediated suppression of infiltrating CD4+ T cells in B-cell non-Hodgkin lymphoma. *Blood* 107: 3639-3646.

[120] Beyer M, Kochanek M, Giese T, Endl E, Weihrauch M. R, et al. (2006) In vivo peripheral expansion of naive CD4+CD25high FoxP3+ regulatory T cells in patients with multiple myeloma. *Blood* 107: 3940-3949.

[121] Curiel T. J, Coukos G, Zou L, Alvarez X, Cheng P, et al. (2004) Specific recruitment of regulatory T cells in ovarian carcinoma fosters immune privilege and predicts reduced survival. *Nat. Med.* 10: 942-949.

[122] Marshall N. A, Christie L. E, Munro L. R, Culligan D. J, Johnston P. W, et al. (2004) Immunosuppressive regulatory T cells are abundant in the reactive lymphocytes of Hodgkin lymphoma. *Blood* 103: 1755-1762.

[123] Farinha P, Al-Tourah A, Gill K, Klasa R, Connors J. M, et al. (2010) The architectural pattern of FOXP3-positive T cells in follicular lymphoma is an independent predictor of survival and histologic transformation. *Blood* 115: 289-295.

[124] Merlo A, Casalini P, Carcangiu M. L, Malventano C, Triulzi T, et al. (2009) FOXP3 expression and overall survival in breast cancer. *J. Clin. Oncol.* 27: 1746-1752.

[125] Sasada T, Kimura M, Yoshida Y, Kanai M, Takabayashi A (2003) CD4+CD25+ regulatory T cells in patients with gastrointestinal malignancies: possible involvement of regulatory T cells in disease progression. *Cancer* 98: 1089-1099.

[126] Giannopoulos K, Kaminska W, Hus I, Dmoszynska A (2012) The frequency of T regulatory cells modulates the survival of multiple myeloma patients: detailed characterisation of immune status in multiple myeloma. *British Journal of Cancer* 106: 546-552.

[127] Mizukami Y, Kono K, Kawaguchi Y, Akaike H, Kamimura K, et al. (2008) CCL17 and CCL22 chemokines within tumor microenvironment are related to accumulation of Foxp3+ regulatory T cells in gastric cancer. *Int. J. Cancer* 122: 2286-2293.

[128] Hontsu S, Yoneyama H, Ueha S, Terashima Y, Kitabatake M, et al. (2004) Visualization of naturally occurring Foxp3+ regulatory T cells in normal and tumor-bearing mice. *Int. Immunopharmacol.* 4: 1785-1793.

[129] Banerjee D. K, Dhodapkar M. V, Matayeva E, Steinman R. M, Dhodapkar K. M (2006) Expansion of FOXP3high regulatory T cells by human dendritic cells (DCs) in vitro and after injection of cytokine-matured DCs in myeloma patients. *Blood* 108: 2655-2661.

[130] Feyler S, Scott G. B, Parrish C, Jarmin S, Evans P, et al. (2012) Tumour Cell Generation of Inducible Regulatory T-Cells in Multiple Myeloma Is Contact-Dependent and Antigen-Presenting Cell-Independent. *PloS one* 7.

[131] Feyler S, von Lilienfeld-Toal M, Jarmin S, Marles L, Rawstron A, et al. (2009) CD4(+)CD25(+)FoxP3(+) regulatory T cells are increased whilst CD3(+)CD4(-)CD8(-)alphabetaTCR(+) Double Negative T cells are decreased in the peripheral blood of patients with multiple myeloma which correlates with disease burden. *Br. J. Haematol.* 144: 686-695.

[132] Ermann J, Hoffmann P, Edinger M, Dutt S, Blankenberg F. G, et al. (2005) Only the CD62L+ subpopulation of CD4+CD25+ regulatory T cells protects from lethal acute GVHD. *Blood* 105: 2220-2226.

[133] Li J, Huang Z. F, Xiong G, Mo H. Y, Qiu F, et al. (2011) Distribution, characterization, and induction of $CD8^+$ regulatory T cells and IL-17-producing $CD8^+$ T cells in nasopharyngeal carcinoma. *J. Transl. Med.* 9: 189.

[134] Muthu Raja K. R, Kovarova L, Hajek R (2012) Functionally active myeloid-derived suppressor cells are increased after lenalidomide plus dexamethasone treatment in previously untreated multiple myeloma patients. *Haematologica* 97(s1):110

[135] Dannull J, Su Z, Rizzieri D, Yang B. K, Coleman D, et al. (2005) Enhancement of vaccine-mediated antitumor immunity in cancer patients after depletion of regulatory T cells. *Journal of Clinical Investigation* 115: 3623-3633.

[136] Mahnke K, Schonfeld K, Fondel S, Ring S, Karakhanova S, et al. (2007) Depletion of CD4(+)CD25(+) human regulatory T cells in vivo: Kinetics of Treg depletion and alterations in immune functions in vivo and in vitro. *International Journal of Cancer* 120: 2723-2733.

[137] Morse M. A, Hobeika A. C, Osada T, Serra D, Niedzwiecki D, et al. (2008) Depletion of human regulatory T cells specifically enhances antigen-specific immune responses to cancer vaccines. *Blood* 112: 610-618.

[138] Prabhala R. H, Neri P, Bae J. E, Tassone P, Shammas M. A, et al. (2006) Dysfunctional T regulatory cells in multiple myeloma. *Blood* 107: 301-304.

[139] Atanackovic D, Cao Y, Luetkens T, Panse J, Faltz C, et al. (2008) CD4+CD25+FOXP3+ T regulatory cells reconstitute and accumulate in the bone marrow of patients with multiple myeloma following allogeneic stem cell transplantation. *Haematologica* 93: 423-430.

[140] Gupta R, Ganeshan P, Hakim M, Verma R, Sharma A, et al. (2011) Significantly reduced regulatory T cell population in patients with untreated multiple myeloma. *Leuk. Res.* 35: 874-878.

[141] Guichelaar T, Emmelot M. E, Rozemuller H, Martini B, Groen R. W, et al. (2013) Human Regulatory T Cells Do Not Suppress the Antitumor Immunity in the Bone Marrow: A Role for Bone Marrow Stromal Cells in Neutralizing Regulatory T Cells. *Clinical Cancer Research* 19: 1467-1475.

In: Flow Cytometry
Editor: Stefanos Papandreou

ISBN: 978-1-62808-709-3

Chapter 3

Multicolor Panel Design Strategies in Rare Event Analysis: Dendritic Cells and Regulatory T Cells

Ilse Van Brussel[1], Nathalie Ruyssers[2], Marthe Heylen[2] and Dorien M. Schrijvers[1]

[1]University of Antwerp, Laboratory of Physiopharmacology, Antwerp, Belgium

[2]University of Antwerp, Laboratory of Gastroenterology, Antwerp, Belgium

Abstract

The need for multidimensional analysis using complex stain combinations has been expanding considerably, especially to discriminate, classify and/or isolate rare cell populations, such as dendritic cells (DCs) and regulatory T cells (Treg). Several principles need to be taken into account when setting up a multicolor flow cytometry experiment for rare event analysis.

The frequency of the event of interest in a sample dictates how many cells must be processed. The lower the frequency of the events of interest, the more events need to be acquired, which might result in large data files. However, the more events one counts, the better the precision will be, as the minimal counting error is equal to the square root of the actual counted events of interest.

It is crucial to understand that reliably counting a rare event is like finding a needle in a haystack. Moreover, the spurious events (``noise``) need to be recognized as well, in order to exclude them, determine what causes them and minimize their occurrence. The signal-to noise ratio can be increased in several ways. Firstly, it is very important to carefully select the reagents to improve the discrimination of different populations in multicolor flow cytometry, which depends on the instrument configuration. Furthermore, the rare event should be positively identified by multiple fluorescence parameters, because the random noise is diluted out by a power function of the number of parameters. Secondly, a critical step in panel design is to match the fluorochromes by brightness according to antigen density. The golden rule is to define the antigens with the lowest expression by the brightest chromophore and vice versa. Thirdly, fluorescence spillover can have a huge impact on the resolution sensitivity of dim markers in multicolor flow cytometry experiments. The more colors you have in your tube, the more spillover you have, resulting in a higher background. Also, if multiple antigens are present on a given cell ("coexpression"), it is better to spread them across as many lasers as possible. Finally, the total noise can be characterized by using an appropriate negative gating control, such as isotype-matched fluorochrome-conjugated antibodies or the fluorescence minus one method.

Here, we will discuss all the principles that need to be taken into consideration for a successful multicolor panel design to analyze rare events. We will also illustrate multicolor panel design strategies by specifying the flow cytometric identification and quantification of two rare populations: DC and Treg.

1. Introduction

Flow cytometry is currently the only technological method for the proper analysis of the complex components of the immune system. It allows us to separately characterize the dozens (or hundreds) of phenotypically and functionally distinct subsets of leukocytes, any of which might be clinically relevant [1]. The need for multidimensional analysis using complex stain combinations has been expanding considerably, especially to discriminate, classify and/or isolate rare cell populations, such as dendritic cells (DC), hematopoietic stem cells, circulating endothelial progenitor cells, antigen-specific T cells and regulatory T cells (Treg). A growing number of flow cytometry-based assays, both for basic science and clinical applications, depend on rare event detection. A rare event is usually defined as an event that is as frequent or less frequent as spurious events, or as an event that occurs at a frequency of 1 in 1.000 (0.1%) or less [2]. Spurious events, at their turn, are defined as all infrequent artifactual flow events that have nothing to do with unlabelled or specifically (fluorescent-) labeled single viable cells [3]. Accurate detection and enumeration of cells occurring at such low frequencies is not standard flow cytometry [4]. It is crucial to understand that reliably counting a rare event is like finding a needle in a haystack. For the accurate and precise measurement of rare events, the frequency of the rare event (= signal) must be increased, whereas the frequencies of the spurious events (= noise) have to be decreased. As a consequence, it is not sufficient to recognize the rare events, the spurious events (``noise``) need to be recognized as well, in order to exclude them during the analysis, determine what causes them and prevent their occurrence. Noise is generally due to a variety of factors that yield fluorescent signals, or signals perceived by the flow cytometer to be fluorescent, and includes cellular debris and dead cells, aspecific binding of fluorochromes and antibodies, cellular autofluorescence, and disruptions in fluidics caused by clogs. The exclusion of spurious events from the analysis requires the need of multiparametric polychromatic flow cytometry. However, design and analysis of multicolor assays becomes more complex and time-consuming as more parameters are included. The more parameters needed, the more compensation issues, technical limitations, or (in)compatibility between the chosen fluorochromes will occur. Thus, it is critical to understand and follow some basic principles when you design a new multicolor panel for a flow cytometric experiment. In this chapter, we will walk you through the principles that we consider as mandatory in order to have a successful panel design. Moreover, this chapter presents two practical examples of rare event problems: detection of mouse Treg in tissue, and characterization and subset analysis of human circulating DC.

2. How to Avoid Background ("Noise")

Background or noise is usually the consequence of multiple factors, including electrical or mechanical perturbations (e.g. sample carryover from previous experiments), cellular autofluorescence, aspecific binding of a fluorochrome or antibody, disruptions in flow (due to clogs or clumps), and spillover from other detectors. Therefore, prior to performing rare event analysis on any flow cytometer, residual cells, particles, and debris of previous experiments must be removed by extensive cleaning and washing, using filtered solutions [5].

Cells with an intrinsic autofluorescence are often cells from cultures or freshly disaggregated tissues. The level of autofluorescence is influenced by the cells' biological and physiological conditions and is also cell type specific [6,7]. For example, autofluorescence of granulocytes is generally higher than that of lymphocytes, because of their high content of granule-associated flavoproteins [7, 8]. Autofluorescence can be discriminated from specific dye fluorescence by its emission spectrum over a broad range of wavelengths [9]. Autofluorescent biomolecules, such as flavonoids or NADPH, are often better excited with short wavelength light, e.g. 488 nm blue-green light provided by Argon-ion lasers [9]. So, if the cell population of interest is intensively autofluorescent, it is recommended to use long excitation wavelengths, e.g. from the red diode laser. If the rare event of interest is not itself highly autofluorescent, the autofluorescent cells can be removed from the analysis by including an unstained fluorescence parameter ("mock/dump" channel), or by including at least one negative marker for the rare event, i.e. an antibody that is known not to react with the rare event of interest. All interfering cells can then be eliminated during the analysis with an inverted ("not/dump") gate [5, 10].

Nonspecific fluorescence can be minimized by a preincubation/blocking step with normal serum prior to staining with the relevant antibodies [9, 5]. Although many different blocking reagents are commercially available, the most appropriate and efficient blocker is heat-inactivated serum from the same species used to create the monoclonal antibody [4]. For example, mouse serum should be used when conjugated mouse anti-human antibodies are included in the analysis.

Dead or dying cells might also contribute to aspecific binding and autofluorescence, or they might have an altered marker expression compared to their viable counterparts, so it is encouraged to exclude them from the analysis as well. This can be done in several ways. Dead cells, subcellular debris and late apoptotic cells are often low light scatter events, which can be eliminated most effectively by creating a positive gate on the population of interest [2]. Also, a number of dead cell markers are available, including DNA-intercalating dyes (e.g propidium iodide (PI), 7-aminoactinomycin D (7-AAD), 4'-6-diamidino-2-phenylindole (DAPI)), phosphatidylserine-binding reagents (annexin V) and amine-reactive dyes (UViD, ViViD, GrViD, OrViD). Dead cells allow permeabilization of these dyes, so they can be recognized by their broad emissions spectra (usually spanning several fluorescence parameters) and bright staining, and can subsequently be gated out [4, 2, 5, 9].

Random (uncorrelated) noise and spikes from photomultiplier tubes (PMTs) and electronics might be excluded by 1) increasing the number of parameters that positively define the rare event, and 2) defining the rare event with the brightest and tightest [lowest coefficient of variation (CV)] positive marking [2, 11].

Finally, fluidic disturbances can alter laminar flow within the flow chamber and results in enhanced variability in measurements, mainly in light scatter. However, transient fluidic disturbances can be easily detected by plotting a time parameter, which can be saved as a listmode parameter or calculated online, versus log side scatter. The disturbing event bursts caused by minor clogs, clumps or other episodic problems can be eliminated from the analysis with a logical gate, if they prove to have altered marker expression [2, 10]. The time parameter can also be used to remove the flurry of spurious events that might occur if a sample tube is running dry or when air is aspirated [10].

3. Reagent Selection

Selection of the appropriate reagents depends on the configuration of the flow cytometer. Some instruments, which only contain a blue (488nm) and a red (635 nm) laser, allow 4- or 6-color assays, whereas other instruments – with an additional violet (405 nm) laser or others - allow measurement of ten or even more colors, and often they can have various selectable configurations. The available lasers and detectors in the instrument determine whether or not a certain fluorochrome can be excited, and whether or not a specific combination of fluorochromes can be read out. Also, the choice of optical filters that are used with each detector dictates the brightness of one fluorochrome versus another. For example, using a wider bandpass filter can enhance the ability to detect a certain fluorochrome, but it may also enhance the amount of spillover from other neighboring fluorochromes. With the introduction of instrumentation capable of measuring up to 20 parameters (two physical parameters (cell size and granularity) and 18 fluorescences), new possibilities for performing more complex analyses have opened up [1]. The most recent generation of flow cytometers and sorters often offer many different instrument configurations, making it impossible to appoint the "most optimal" fluorochromes for use in 6, 8 or more color combinations.

Once you know which fluorochromes can be used in a specific instrument, the fluorochromes must be ranked according to their brightness. The brightness of the fluorochromes influences the ability to discriminate a dim positive signal from background (= resolution sensitivity). A good measure of resolution sensitivity is the stain index (SI = D/W), an algorithm that takes into account data spread (W = 2 x standard deviation of the negative peak) and the difference in positive and negative population medians (D) [12].

Once you have an idea of the relative brightness of different fluorochromes on a specific instrument, the fluorochromes must be matched by brightness according to antigen density and distribution. This is a critical step in multicolor panel design, it forms the key of success for the assay. The golden rule is to use the brightest fluorochromes for the least expressed proteins and the dimmest fluorochromes for the most highly expressed antigens. To achieve this, the antigens that need to be included in the analysis must be classified into three groups [1]: 1) *primary antigens*, which are well characterized and can easily be classified as positive ("on" expression) or negative ("off" expression), e.g. CD3, CD4, CD8, CD14, CD19, CD20;

2) *secondary antigens*, which are also well characterized, but their expression pattern is often rather a continuum, e.g. CD27, CD28, CD45RA/RO; and 3) *tertiary antigens*, which include antigens that are expressed at low levels, e.g. CD25, or uncharacterized antigens. Next, bright fluorochromes, such as PE and APC, are preferentially chosen for weak antigens where only low numbers of molecules are expressed per cell or those that are expressed by only a very small fraction of cells (tertiary antigens) [1]. The expression level of the antigen might be found in literature, however, these data must be interpreted with much caution, because the absolute number might not tell the full story. The antigen density can vary due to cell activation, or can be a range of expression (~ smeared populations). For example, the T cell differentiation marker CD28 is highly expressed on some T cell subsets, but only at low levels on other T cells. If no information about the expression level of the antigen is available, the technical datasheet of the antibody must be examined, because it often contains a histogram showing information about the expression level of the marker.

4. Limiting Spectral Overlap

As soon as multiple reagents are run together, spectral overlap – or spillover – becomes an issue. Spillover from a given fluorochrome into another channel can directly and irreversibly reduce the resolution sensitivity of that other channel. Choosing fluorochromes which have the least spectral overlap will enhance the quality of your results significantly. Generally, the further apart the maximal emission, the less spillover will occur. Typically, associations such as APC and APC-Cy7 as well as PE-Cy5 and APC must be avoided if possible because they generate a large amount of spectral overlap. This might seem contradictory with the above mentioned recommendation, which was to choose the most brightly fluorochromes as possible for the analysis. Indeed, you might have to sacrifice a certain amount of brightness in one detector to avoid spillover in another.

'Compensation' needs to be applied to deal with spillover. It is the process by which the fluorescence "spillover" originating from a fluorochrome other than the one specified for a particular PMT detector is subtracted as a percentage of the signal from other PMTs. Any flow cytometry experiment that makes use of more than one color needs to include samples labeled with single antibody conjugates incorporating the relevant fluorochromes for compensation [1]. The net result of compensation is that populations with equivalent fluorescence in a given channel will have similar means (or medians), regardless of their fluorescence in other channels [13]. Compensation between detectors can be performed either by hardware, after signal detection but before logarithmic conversion and/or digitization, or post-collection by software [14].

The use of samples labeled with single antibody conjugates as compensation controls is commonly applied and works well in most situations. However, this approach requires that either extra cells from the test samples are used for compensation controls, or that another source of cells must be obtained. Although it is necessary to use the same fluorochrome for compensation as used in the samples (and not just one of all possible fluorochromes for a given detector), it is not necessary to use the same antibody. This facilitates compensation for markers that are dimly expressed or that are only found on rare subsets of cells. Often antibodies which are very bright and present on a reasonable subset of cells, such as CD3 or

CD8, are routinely used for compensation in each channel, regardless of what antibody will be used experimentally in that channel [13]. However, substitution of a different antibody for compensation than the one that is present in the test samples might be risky for certain tandem dyes, especially APC-Cy7 and PE-Cy7 [13]. These tandem fluorochromes tend to have more lot-to-lot variability in their emission spectra, and their spillover is also affected by handling and storage of the antibody and stained samples [13].

Compensation beads are provided by research companies and can be used to deal with spillover instead of precious sample cells. The use of compensation beads is encouraged when cell numbers are limited, the antigen is dimly expressed, the positive population is rare or when tandem dyes are used. Compensation bead kits always contain two populations: uncoated beads that will not react with antibody and as such represent the negative population, and coated beads that bind antibody and thus represent the positive population. When a fluorochrome-conjugated antibody is added to the beads both positive and negative populations are formed. This bimodal distribution can then be used for single-color compensation controls in multicolor flow cytometry experiments.

5. Technical Limitations of Tandem Dyes

A tandem dye is composed of two covalently attached fluorescent molecules, typically a large protein-based dye (the donor or parent) and a smaller organic dye (the acceptor), that predominantly emit light characteristics of the more red emitting dye through the phenomenon of fluorescence resonance energy transfer (FRET). A high FRET efficiency will result in a strong signal in the acceptor channel and weaker signal in the donor channel. A low FRET efficiency will lead to a strong signal in the donor channel and a weak signal in the acceptor channel. Because the FRET efficiency will never reach 100%, there will always be some fluorescence signal detected into the channel of the donor dye.

Tandem dyes have been used in flow cytometry for decades and have allowed to investigate progressively larger numbers of markers in a single experiment. However, instability and variability of tandem dye conjugated products are well recognized problems [15]. Certain tandem dyes, especially APC-Cy7 and PE-Cy7 [15], are very sensitive to – and can degrade with - exposure to light, temperature and fixation. Tandem dye degradation results in the emission of light in the donor dye detector (APC or PE), leading to false positive events in APC or PE [12]. Thus, it is recommended to avoid using APC or PE for detection of rare events on a population of cells positive for APC-Cy7 or PE-Cy7 respectively [12]. Yet, the problem of false positive fluorescence in the donor fluorochrome channel can be largely overcome by minimizing the exposure of samples to light, heat or formaldehyde-based fixative at all times. So, if you plan to perform a multicolor scenario including the use of tandem dyes, try to limit the amount, the concentration and the time that you are fixing the cells. For example, use 1% paraformaldehyde for 30 minutes, wash and leave your cells in wash buffer until acquisition. This would help these tandem dyes to be in a better environment than when continuously exposed to fixative. PE-Cy5, PerCP-Cy5.5, APC-H7 and the new tandem PE-CF594 remain stable when exposed to light.

Besides handling procedures, the manufacturing process and lot number are considered the key aspects affecting the fluorescence characteristics of tandem dyes [15]. It is important to remember that each lot of a tandem dye will have different spectral characteristics. In general, compensation will be different for two different PE-Cy5 reagents [16]. As mentioned before, compensation for the tandem dye should be set using the same antibody-conjugate as the one that will be used in the experiment.

6. Appropriate Controls to Validate Your Experiment

For optimal interpretation of the test samples, every multiparametric flow cytometric experiment must contain at least 3 types of controls: 1) instrument setup controls, which are used to properly set up the flow cytometer, including PMT voltage gains and compensation; 2) gating controls, which help to distinguish specific from nonspecific binding; and 3) biological controls, which provide biologically relevant comparison conditions [13]. The higher the number of fluorochromes and antibodies used in each stain the greater the risk for artifacts introduced by compensation errors and/or reagent interactions.

The instrument setup controls should be performed with every "cold start" of the instrument and aims to identify both immediate and potential problems [17]. For this, appropriate reference microbeads are used. The results of running these reference beads are then compared with established tolerance limits, and variations can be noted with the use of Levey-Jennings style charts that allow visual inspection of the longitudinal data for monitoring instrument performance for precision and trends. If any value falls outside the tolerance limits, monitoring should be repeated and instrument settings adjusted if problems persist [17].

Gating controls, including isotype controls or fluorescence minus one (FMO) controls, help to set positive/negative boundaries in the data plots and can be of great importance for the interpretation of the data [13]. If a marker has a clear bimodal expression without overlap between positive and negative populations (e.g. CD4 or CD8 on T cells), a gate can be drawn unambiguously and does not require a gating control. However, when there is no clear division between positive and negative populations (e.g. for activation markers, such as CD25, CD86, CD40, CD69), gating controls become important [13]. Isotype controls give an idea of nonspecific staining of an antibody of a particular isotype conjugated to a particular fluorochrome [18]. Such nonspecific binding is usually, but not exclusively, mediated by receptors that bind the Fc portion of the various immunoglobulin subclasses [19, 20]. Yet, there are some limitations to the usefulness of this type of control. The fluorescent dye/protein (F/P) ratio as well as the concentration of the isotype should be exactly the same as for the test antibody.

Most manufacturers, however, aim to minimize inconsistencies by designing the fluorochrome-conjugated isotype control antibodies within a standard range of F/P ratio that is similar to the primary antibodies. But even in the ideal situation of matched isotype and subclass, matched F/P ratio and concentration, there may still be differences in background binding characteristics of the negative controls [18]. These limitations are inherent to the use of isotype controls.

Nevertheless, they can be particularly useful in certain situations to make an estimation of the background or baseline fluorescence, to assess the up or downregulation of a specific surface marker after a particular *in vitro* manipulation, or to assess the extent to which a cell activation procedure alone might generate an increase in nonspecific binding [18]. Some investigators believe that isotype controls are fairly crucial in rare event analysis, as they provide a good estimate of nonspecific binding of antibodies to cells [4]. Even with freshly drawn peripheral blood, it is not unusual to observe nonspecific binding of isotype controls ranging from 0.1% to 0.5% of cells analyzed. In most clinical assays, this level of aspecific staining does not significantly affect reportable data, but in staining for very rare events, this might represent a background higher than the anticipated frequency of specific events [4]. However, Keeney and colleagues [19] stated that isotype control monoclonal antibodies (mAb) are inappropriate in the setting of rare event detection, e.g. the enumeration of $CD34^+$ stem cells. After all, the $CD34^+$ antibody will often not only stain $CD34^+$ cells specifically, but also other events nonspecifically. The isotype does not stain exactly the same number of events that are nonspecifically stained by the CD34 antibody [19]. Imagine that the isotype control detects lower aspecific binding than the CD34 mAb, its use to set the lower boundary for $CD34^+$ cells leads to an overestimation of the number of $CD34^+$ cells in a sample [19].

In multicolor experiments – if high-quality mAb conjugates are used at appropriate concentrations – the major source of background staining tend to be fluorescence spillover [13]. Therefore, using FMO controls might be more appropriate than using isotype controls. FMO controls are samples that include all of the antibody conjugates present in the test sample except the one of interest. The channel in which the antibody conjugate is missing is the one for which the FMO provides a gating control [13, 21]. FMO controls provide a way to measure spillover effects from populations in other dye dimensions on a particular channel of interest. However, as it does not contain an antibody in the channel of interest, an FMO control does not provide a measure of background staining that may be present when an antibody is actually included in that channel [13]. Consequently, as soon as antibody-dependent background staining becomes significant, the FMO control will no longer be appropriate to determine a positive/negative boundary.

Besides isotype or FMO controls, there are biological controls that might be more appropriate in some cases for proper setting of positive/negative boundaries. For example, in stimulation assays, the unstimulated sample can provide the best way to discriminate between positive and negative events, except when high background in the unstimulated sample makes it more difficult to set a clear positive/negative boundary [13]. The unstimulated sample accounts for both 1) spillover effects on the fluorescence channel of interest by including all of the antibody conjugates present in the other test samples (similar to an FMO control), and for 2) nonspecific staining in the fluorescence channel of interest (similar to an isotype control) [13].

Also, when comparing a disease state with a healthy individual, a sample of each condition must be run in every experiment. Furthermore, positive biological controls, e.g. a set of peptides with high reactivity among a broad spectrum of individuals (such as a CMV peptide pool) might also be of high importance to validate the potential finding of negative results in the other test samples. A positive control ensures that the experiment was performed in such a way that positive responses could be detected [13].

7. How Many Events Must be Processed?

Event frequency is a property of the sample and not something that can be manipulated, unless by a pre-enrichment step prior to analysis [10, 22]. Immunomagnetic separation can be preferable for rare event isolation, as it can enormously decrease the time required to sort a rare population, but the uncertainties of event recovery after preseparation may compromise the true frequency of the event of interest [2]. The frequency of the event of interest in the sample is one of the parameters that dictates how many cells must be processed in total [10]. Obviously, in rare event analysis, large numbers of cells must be counted in order to obtain statistically meaningful numbers of rare cells [4], which might result in extremely large data files. However, the more events one counts, the better the precision will be, as the minimal counting error is equal to the square root of the actual counted events of interest.

The distribution of any subset of events within a larger parent population will be determined by a binomial distribution: an event (or cell) will either meet a predefined criterion or not (positive or negative, 0 or 1, true or false). However, when small subsets are considered – especially when populations of less than 5% are being analyzed – Poisson statistics must be used [23], as they deal with the probability distribution of rare events. The Poisson CV of the counting error, defined as 100/√positive events counted, is 10% when 100 events are counted [9]. Imagine a rare population of interest that represents only 0.04% of the total population. As stated above, to achieve a CV of 10%, 100 events must be counted in the rare event gate [# positive events counted = $(100/CV)^2$]. This means that at least 250.000 events [=100/0.0004] must be acquired in total. To achieve a CV of 5%, 400 events must be counted in the rare event gate. Thus, 1.000.000 events [= 400/0.0004] must be recorded in total.

Recording one million events in total might result in long acquisition times. However, the sample acquisition rate on the flow cytometer can be enhanced in two ways: 1) by concentrating the sample, or 2) by increasing the sample flow rate [2, 10]. Most analytical instruments have different flow rate settings, usually low, medium and high, that corresponds to the rate (microliters per minute) at which the sample is drawn into the cytometer. Increasing the flow rate is associated with increased sample pressure relative to the sheath pressure, which results in a wider sample stream, and consequently increasing the probability that two cells pass through the laser beam and the detection system at the same time (i.e. coincidence) [2, 10]. As a result, the resolution of the measurement will be lower. Furthermore, the CV of all measured parameters will be higher, since the position of an individual cell within the stream during interrogation is more variable [2, 10]. Thus, in rare event analysis, it is recommended to optimize the sample concentration in order to maximize the event rate, instead of increasing the flow rate.

The tremendous amount of data resulting from a rare event flow cytometry experiment has led to some confusion about the (lack of) statistical significance in the precision of the subset frequency measurements [24]. Indeed, it might be tempting to conclude that finding a single event in a gate when a total number of one million events is required, does not mean anything. Does the event represent what the researcher claims it is or is it rather noise of some sort or experimental background?

Unfortunately, an answer to that question cannot be formulated, as there is no way to distinguish between the possibilities [24]. To find out whether a sample is positive or not, the distribution of negative events must first be determined by using control samples (vide supra) [24].

8. Multicolor Panel Design to Detect Rare Events: Two Examples

Example 1 - Analysis of Human Circulating Dendritic Cell (DC) Subsets

Although DCs represent only a small population of leukocytes, they are the main conductors of the immune symphony as they initiate and regulate immune responses. These responses can be either immunogenic or tolerogenic, depending on the DC lineage, differentiation and activation state [25, 26]. As sentinel members of the innate immune system, DCs respond to antigens and molecules containing pathogen-associated molecular patterns (PAMPs) or damage-associated molecular patterns (DAMPs)—so-called danger signals—by the generation of protective cytokines [27, 28]. Being part of the acquired immune system, DCs respond to these harmful molecules by efficient antigen uptake, processing, and presentation, and hence DCs are crucial in the initiation of adaptive immune responses. Besides their potent capacity to stimulate naive T cells, effector T cells and memory T cells, as well as B cells, they are also involved in the maintenance of tolerance against harmless (auto)antigens [26, 28, 29].

Hematopoietic DCs originate from precursors of both myeloid and lymphoid lineages that circulate in the blood and typically lack cell surface markers for T, B, natural killer (NK), or monocyte/macrophage cell lineages [10, 30, 31]. Besides differentiation from CD34$^+$ hematopoietic progenitors, DCs can also be derived from blood monocytes, and these monocytic (mo)DCs are characterized by FCγRIII (i.e. CD16) expression [32, 33].

Because DCs are present in the circulation at very low frequencies, they have been difficult to study [30]. However, the identification of DC subsets in peripheral blood and various tissues may be crucial to understand their immunologic significance. Further monitoring and characterization of the different subsets might provide insights into the role of DC subsets in several physiologic and disease processes that involve polarization of the immune system [10]. Furthermore, in humans, the only tissue that may be readily accessible for analysis of these rare DC subsets is peripheral blood [31].

In this example, we have performed rare event, 10-parameter, flow cytometric analysis of peripheral blood to identify the presence of several rare DC subset populations in the circulation of a normal healthy volunteer, including plasmacytoid (p)DCs (CD123$^+$), Th1/Th17-inducing DCs (CD11b$^+$), crosspresenting DCs (Clec9A$^+$) and moDCs (CD16$^+$). The same strategies can be applied to high speed fluorescence activated cell sorting (FACS) [10]. As an alternative, immunomagnetic lineage depletion can be performed as a first step [10].

Peripheral blood mononuclear cells (PBMC) were obtained from a healthy subject by standard Ficoll (MP Biomedicals, Illkirch, France) density gradient centrifugation. Nonspecific fluorescence, which is mainly mediated by receptors that bind the Fc portion of the various immunoglobulin subclasses, was limited by blocking Fcγ receptors with mouse immunoglobulins (BD Biosciences, Erembodegem, Belgium) for 10 minutes in the dark at room temperature. Then, cells were stained with mouse anti-human mAbs (all from BD Biosciences, except Clec9A which was from R&D Systems, Abingdon, UK) directed against cell surface markers (Table 1) for 25 min at 4°C. Staining conditions for each mAb were preliminary determined in titration assays. Although in this experiment freshly isolated peripheral blood was used, which usually does not result in a lot of autofluorescence, we used multiple parameters that positively define the rare DCs (including CD45, CD11c, HLA-DR) as well as a cocktail of lineage markers (including CD3, CD14, CD19, CD20, CD56) which are negative markers for the DCs.

Table 1. List of monoclonal antibodies used in example 1

Marker	Description	Conjugate	Clone	Company
lineage cocktail 2	CD3 (T lymphocytes), CD14 (monocytes, macrophages, neutrophils, eosinophils), CD19 and CD20 (B lymphocytes) and CD56 (NK cells).	FITC	SK7, SJ25C1, L27, MφP9, NCAM16.2	BD Biosciences
HLA-DR	human class II major histocompatibility complex (MHC) antigen	BD Horizon V500	G46-6	BD Biosciences
CD11c	Integrin αx chain	PE-CF594	B-ly6	BD Biosciences
CD11b	Integrin αM chain	Brilliant Violet 421	M1/70	BD Biosciences
CD45	Leukocyte common antigen	APC-H7	2D1	BD Biosciences
CD16	FcγRIII	PerCP-Cy5.5	3G8	BD Biosciences
CD123	IL-3Rα	PE	9F5	BD Biosciences
Clec9A	C-type lectin-like receptor 9	APC	683409	R&D Systems

Table 2. Configuration of the FACSAria II instrument

Detectors	Parameter	Fluorochrome Examples	Laser (nm)	Filter centre	Band width (nm)
D 1	FSC - 488	-	488	-	-
D 2	FSC - 638	-	635	-	-
D 3	SSC	-	488	-	-
D 4	FL1	FITC, AlexaFluor-488	488	527 band pass	30
D 5	FL2	PE	488	575 band pass	26
D 6	FL3	PE-TexasRed, PE-CF594	488	630 band pass	22
D 7	FL4	PE-Cy7	488	748 long pass	-
D 8	FL5	PE-Cy5, PerCP-Cy5.5	488	682 band pass	22
D 9	FL6	PE-Cy5.5	488	710 band pass	20
D 10	FL7	Pacific Blue, BV421	405	450 band pass	50
D 11	FL8	405-green, BD Horizon V500, BV510	405	527band pass	30
D 12	FL9	APC, AlexaFluor-647	635	675 band pass	20
D 13	FL10	Cascade orange	405	575 band pass	26
D 14	FL11	APC-Cy5.5	635	710 band pass	20
D 15	FL12	APC-Cy7, APC-H7	635	748 long pass	-

Abbreviations: FSC: forward scatter; SSC: side scatter; FL: fluorescence detector; BV: Brilliant Violet

Table 3. Matching fluorochromes by brightness according to antigen density - example 1

	Fluorochromes	Antigens		Matches
Brightest	Brilliant Violet 421	high	CD45	**CD45-APC-H7**
	PE		CD11b	**CD11b-Brilliant Violet 421**
	PE-CF594		HLA-DR	**HLA-DR-BD Horizon V500**
	PE-Cy5		Lineage	**Lineage-FITC**
	APC		CD11c	**CD11c-PE-CF594**
Bright	PE-Cy7		CD16	**CD16-PerCP-Cy5.5**
	AlexaFluor-647	↓	CD123	**CD123-PE**
	PerCP-Cy5.5	low	Clec9A	**Clec9A-APC**
Moderate	AlexaFluor-488			
	FITC			
	Pacific Blue			
Dim	PerCP			
	APC-Cy7			
	BD Horizon V500			
	APC-H7			

Ten-parameter (forward scatter, side scatter, 8 fluorescences) flow cytometry was performed with a FACSAria II (Beckton Dickinson, Erembodegem, Belgium) and analyzed using FACSDiva software. The FACSAria II contains three lasers (a blue 488-nm laser, a red 635-nm laser and a violet 405-nm laser) and 15 detectors (2 for FSC, 1 for SSC and 12 for fluorescence). The configuration of the instrument is presented in detail in table 2. The latter also reveals which fluorochromes can be measured on the instrument. To match the fluorochromes by brightness according to the antigen density, we ranked the fluorochromes from bright to dim (based on the SIs) and the antigens from high expression to low expression (Table 3).

Spectral overlap between the different fluorochromes used was limited by including samples labeled with single antibody conjugates incorporating the relevant fluorochromes for compensation. Before running the samples for compensation, a performance check was run to ensure that the cytometer performed consistently over time. The performance check of the FACSAria must be run daily using Cytometer Setup and Tracking (CS&T) beads, which consist of bright, mid, and dim beads labeled with a mixture of fluorochromes. After verifying that the performance check has passed, PMT voltages were set prior to compensation using an unstained sample of peripheral blood cells. The voltages were adjusted such that the unstained cells appear on the left (in the first decade or first quartile) of a 5-decade logarithmic scale for each fluorochrome to be measured (figure 1A). It is crucial to determine the required minimal voltage in such a way that 1) each detector has enough gain applied to sufficiently boost dim signals above a level where underlying electronic noise significantly contributes to the measurement [13], and 2) all events are brought on-scale. So, if the signal of a positive population is so high that it is off-scale in a particular detector, the PMT voltage must be reduced. This was inspected by using the samples labeled with single antibody conjugates (figure 1B). All the single antibody conjugates contained the same antibody as the one used in the experiment, except for PE and APC, which were coupled to the very bright and routinely used marker CD8. After measuring all single antibody conjugates, compensation was automatically calculated by the instrument, resulting in the compensation matrix presented in table 4.

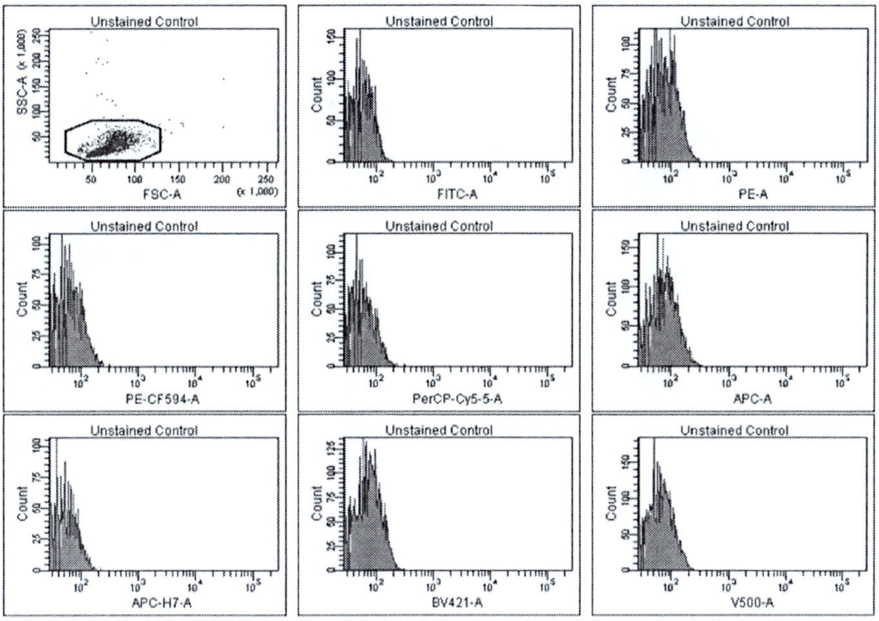

(A)

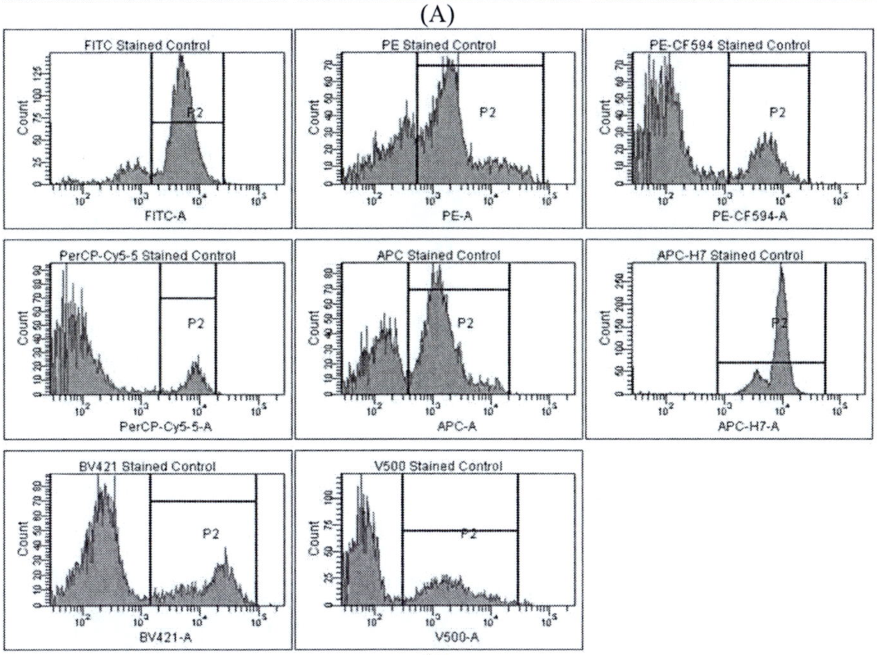

(B)

Figure 1. A. *Setting PMT voltages using an unstained sample prior to compensation.* The voltages were adjusted such that the unstained cells appear on the left (in the first decade or first quartile) of a 5 decade logarithmic scale for each fluorochrome to be measured B. *Adjusting PMT voltages in such a way that all events (positive and negative populations) are on-scale.* It is crucial to determine the required minimal voltage in such a way to 1) ensure that each detector has enough gain applied to sufficiently boost dim signals above a level where underlying electronic noise significantly contributes to the measurement, and 2) bring all events on-scale. So, if the signal of a positive population is so high that it is off-scale in a particular detector, the PMT voltage must be reduced. This was inspected by using the samples labeled with single antibody conjugates.

Table 4. Compensation matrix for the 8-color panel

	FITC	PE	PE-CF594	PerCP-Cy5.5	APC	APC-H7	BV421	V500
FITC		43.29	11.01	1.77	0.00	0.03	2.72	0.00
PE	1.10		27.45	5.49	0.00	0.07	0.00	0.00
PE-CF594	0.40	30.99		29.62	1.53	0.23	1.09	1.27
PerCP-Cy5.5	0.02	0.00	0.16		3.17	10.54	0.20	0.18
APC	0.00	0.00	0.00	0.00		16.76	0.00	0.00
APC-H7	0.00	0.02	0.06	0.00	1.85		0.02	0.05
BV421	0.06	0.15	0.09	0.03	0.18	0.03		4.62
V500	4.8	4.62	1.90	0.36	1.93	0.08	11.77	

Abbreviations: BV: Brilliant Violet.

Next, the experimental test sample containing the 8-color panel was run on the instrument. A sample containing the appropriate isotypes for all markers was used to get an idea of the nonspecific binding of the antibodies and to set the positive/negative gates. The gating strategy to identify the different circulating DC subsets is presented in figure 2. Dead cells, subcellular debris and late apoptotic cells – which are all low light scatter events – are eliminated by the creation of a positive gate on PBMCs and by using a FSC threshold. At least 4 x 10^6 events were collected (both for the isotype control and the test sample) on low flow rate at a maximum of 5.000 events/sec.

All DC subsets were identified at low frequencies in the PBMC fraction of the healthy control subject: $CD16^+$ moDCs as well as Th1/Th17-polarizing cDCs (~ BDCA-1^+ cDCs) constituted both about 1% of the PBMC population, $Clec9A^+$ crosspresenting cDCs (~ BDCA-3^+ cDCs) constituted about 0.01% and $CD123^+$ pDCs (~ BDCA2/4^+ DCs) about 0.3%. Off course, by performing a Ficoll density gradient centrifugation, we did not look to their frequency in whole blood, which would have resulted in even lower percentages. Recording at least 4 x 10^6 events in total resulted in a CV of 5.6% for the most rare population (= $Clec9A^+$ cDCs).

Example 2 -Analysis of Mouse $CD4^+$ $FoxP3^+$ Regulatory T Cells in Colonic Tissue

Regulatory T cells (Treg) play an important role in the peripheral tolerance of T cells towards endogenous materials. Their principal action is to regulate the immunological equilibrium in order to maintain the body's homeostasis. Therefore their function is to avoid excessive effector T cell activation and tissue damage during normal protective responses against infections and to prevent inappropriate reactions against self-antigens [34]. Two major classes of Treg cell populations have been identified: the naturally occurring Treg cells (nTreg) and the inducible Treg cells (iTreg). The nTreg cells are formed in the thymus and their share in the total peripheral $CD4^+$ Th cell population is 5-10% [35]. The nTreg cells are characterized by the membrane molecules CD4 and CD25 and the transcription factor forkhead box P3 (Foxp3) of the forkhead family [36].

The iTreg cells originate in the periphery from naive $CD8^+$ $CD25^-$ $Foxp3^-$ or naive $CD4^+$ $CD25^-$ $Foxp3^-$ T cells. When an antigen is present, the naive T cells can differentiate into 1)

$CD8^+$ Treg, which mainly produce IL-10, 2) $CD4^+$ $CD25^+$ T helper 3 cells (Th3), which mainly produce TGF-β, or 3) $CD4^+$ $CD25^+$ regulatory T cells type 1 (Tr1), which mainly produce IL-10. Via these cytokines the action of T effector cell subpopulations is inhibited. While nTreg cells rather play a permanent part in maintaining homeostasis, iTreg cells are very important in controlling an induced immune response [35,37]. The iTreg cells are known to maintain plasticity enabling them to convert to other T cell subtypes such as Th1 and Th17 cells [38].

During inflammatory conditions T cell subsets seem to be accumulating in the inflamed tissues as such performing a local effect. Therefore, in addition to examining the circulating pool of T cells in the peripheral blood we consider it of utmost importance to also investigate the T cell balance in the inflamed tissues thereby providing evidence for a local regulatory effect of Treg cells.

To distinguish between the different T cell subsets multiple parameters need to be taken into account. Subtypes can be identified through flow cytometry using a combination of cell surface markers, transcription factors and cytokines [39].

In this example we propose 6-parameter flow cytometric analysis to identify the presence of Treg cells in the colon tissue of mice.

Colonic lamina propria mononuclear cells (LPMC) were isolated as previously described [40]. Mice were anaesthetized, a laparotomy was performed and mice were sacrificed by exsanguination.

The colon was subsequently resected. Colonic tissue was opened longitudinally, cut into 5mm pieces and incubated in 0.5mM EDTA in calcium- and magnesium-free Hank's balanced salt solution (HBSS) for 20 min at 37°C.

This was repeated after thorough washing. Tissue was then incubated for 20 min at 37°C in 20 ml RPMI containing 10 % FCS, 25mM HEPES buffer, 2mM L-glutamine, 50 μM β-mercaptoethanol, 1mM sodium pyruvate, 100 U/ml penicillin and 100 mg/ml streptomycin as well as 1 mg/ml collagenase. At the end of the incubation, the tissue was subjected to further mechanical disruption using a 1 ml syringe. To remove debris, the cell suspension was washed through a pre-wetted gauze, layered in a funnel, with RPMI.

Then, the LPMC were washed once and sieved through a pre-wetted 2cm nylon wool column. After washing, cells were layered onto a 30:70 % gradient Percoll column. Cells were spun at 2200 x *g* for 20 min.

Two different staining protocols were performed in order to investigate sufficient relevant markers to characterize the different T cell subsets. Six-parameter (forward scatter, side scatter, 4 fluorescences) flow cytometry was performed with the BD Accuri C6 (BD Biosciences, Erembodegem, Belgium) and data was analyzed using BD Accuri C6 software. The BD Accuri C6 contains two lasers (a blue 488-nm argon-ion laser and a red 635-nm diode laser) and 6 detectors (1 for FSC, 1 for SSC and 4 for fluorescence). In a first protocol, cells were stained with rat anti-mouse mAbs (BD Biosciences) directed against cell surface markers CD3, CD4, and CD8. A second staining protocol included the cell surface marker CD4 and the intracellular Foxp3 to identify Treg.

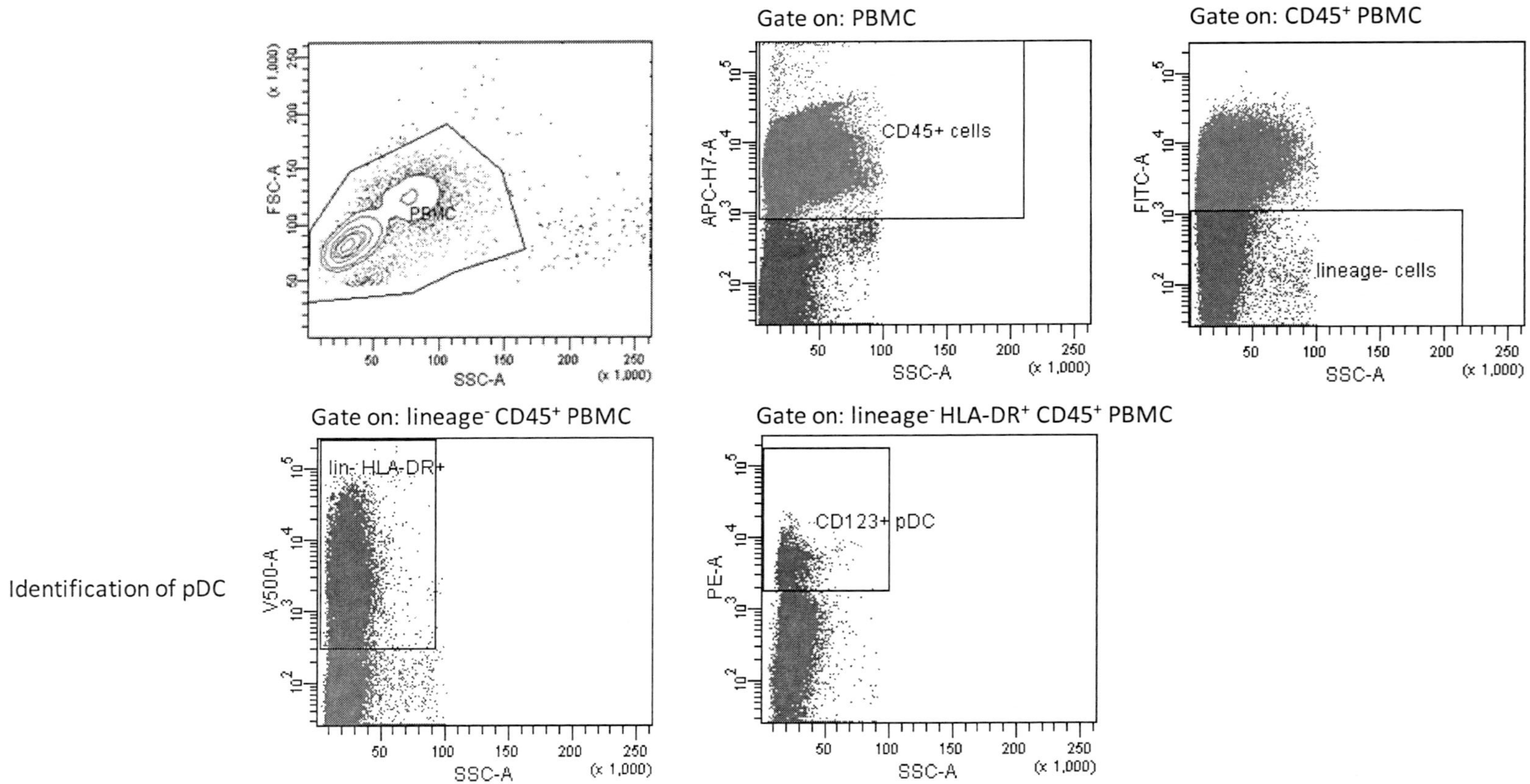

Figure 2. (Continued)

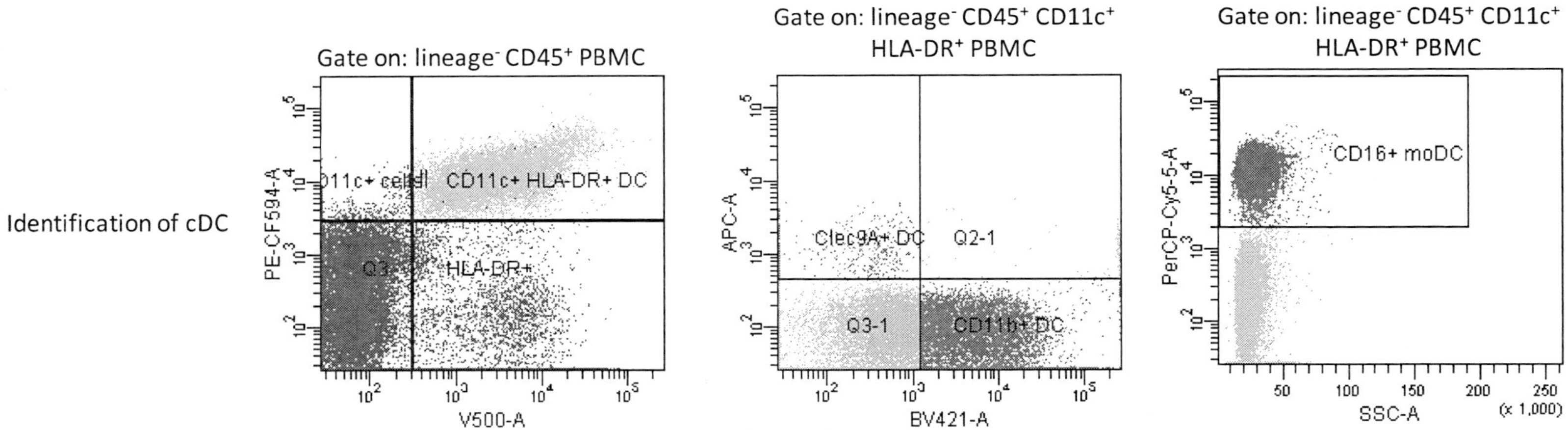

Figure 2. *Gating strategy to identify circulating DC subsets*. Peripheral blood mononuclear cells (PBMC) are identified on a forward scatter area (FSC-A) vs. the side scatter area (SSC-A) bivariate plot. A FSC threshold was set to exclude small (dead) cells, cellular debris and late apoptotic cells. Hematopoietic cells were defined among PBMC as brightly stained with CD45-APC-H7. Among hematopoietic cells, lineage-negative cells were selected. Lineage-negative hematopoietic cells were further divided into HLA-DR$^+$ cells and CD11c$^+$ HLA-DR$^+$ cells. Based on their expression of CD123, pDCs were identified, whereas cDCs were classified based on their expression of Clec9A, CD11b or CD16 to differentiate between BDCA-1$^+$ cDC, BDCA-3$^+$ cDC and moDCs.

Table 5. Matching fluorochromes by brightness according to antigen density - example 2

Fluorochromes		Antigens		Matches
bright	PE	High	CD3	**CD3-FITC**
	PE-Cy5	↓		
	APC		CD4	**CD4-PerCPCy5.5**
	PE-Cy7			
	PerCP-Cy5.5		CD8	**CD8-PE**
dim	FITC			
	APC-Cy7	Low	Foxp3	**Foxp3-APC**

Nonspecific fluorescence, which is mainly mediated by receptors that bind the Fc portion of the various immunoglobulin subclasses, was limited by blocking Fcγ receptors with rat serum for 10 minutes at 4°C. Then, cells were stained with rat anti-mouse mAbs (all from BD Biosciences) directed against cell surface markers for 20 min at 4°C. Currently, Foxp3 still remains an important specific marker for Treg cells. As it is localized intracellularly, permeabilisation of cells needed to be performed. After staining for the cell surface markers, cells were treated with the Foxp3 fix/perm buffer (BD Biosciences) as to the manufacturer's instructions and were subsequently stained with anti-Foxp3 antibody for 45 min at 4°C.

In comparison to the FACSAria used in the previous example, the fluorochromes that can be read on the BD Accuri C6 are limited. To match the fluorochromes by brightness according to the antigen density, we ranked the fluorochromes from bright to dim (based on the SIs) and the antigens from high expression to low expression (Table 5).

Spectral overlap between the different fluorochromes was limited by including samples labeled with single antibody conjugated tothe relevant fluorochromes for compensation. Before running the samples for compensation, a performance check was run to ensure that the cytometer performed consistently over time. The performance check of the Accuri C6 must be run daily using 6-peak and 8-peak beads, which consist of bright, mid, and dim beads labeled with a mixture of fluorochromes. On the Accuri C6 PMT voltage don't need to be set as the flow cytometer plots all cells onto the graph. This is different from the FACSAria as explained above. All the single antibody conjugates contained the same antibody as the one used in the experiment. After measuring all single antibody conjugates, compensation was set manually.

Next, the experimental test sample containing the first 3-color panel (FITC-conjugated CD3, PerCP-Cy5.5-conjugated CD4 and PE-conjugated CD8) was run on the instrument. For further investigation of Treg cells, a second 2-color panel (PerCP-Cy5.5-conjugated CD4 and AlexaFluor647-conjugated FoxP3) was run. For both panels, a sample containing the appropriate isotype controls for all markers was used to determine the nonspecific binding of the antibodies. The percentage of nonspecific staining was subtracted from the positivity in the test samples. We collected 100,000 events (both for isotype controls and test samples). The gating strategy to identify the different T cell subsets is presented in figure 3. Two independent preliminary experiments were performed to identify Treg cells in mouse colon. Data are presented here as mean ± SEM. The lymphocytes within the LPMC population were identified based on low forward and side scatter values and consisted about 13.1 ± 0.9% of all LPMC (Figure 3A). Dead cells and subcellular debris – which are even lower forward light scatter events – were eliminated by using a FSC threshold. Within the lymphocyte gate, 30.7 ± 5.5% were $CD3^+$ cells (Figure 3B). In addition we determined CD4 and CD8 subsets within the CD3^{+} T cell population. Here, we found that 51.7 ± 3.7% of CD3^{+} T cells are in fact CD4^{+} helper cells (Figure 3C), whereas 27.8 ± 4.9% are CD8 cytotoxic T cells (Figure 3D). After measuring the second tube, we again gated the lymphocyte population within the total LPMC (14.2 ± 0.2%) based on their low forward and side scatter (Figure 3E). CD4^{+} T helper cells consisted 18.5 ± 2.4% of the total lymphocyte population (Figure 3F). Subsequently, within the CD4^{+} T helper cell population we characterized Treg cells as CD4^{+} FoxP3^{+}. As such, we found that 16.8 ± 1.7% of CD4^{+} T cells are Foxp3^{+} Treg cells (Figure 3G).

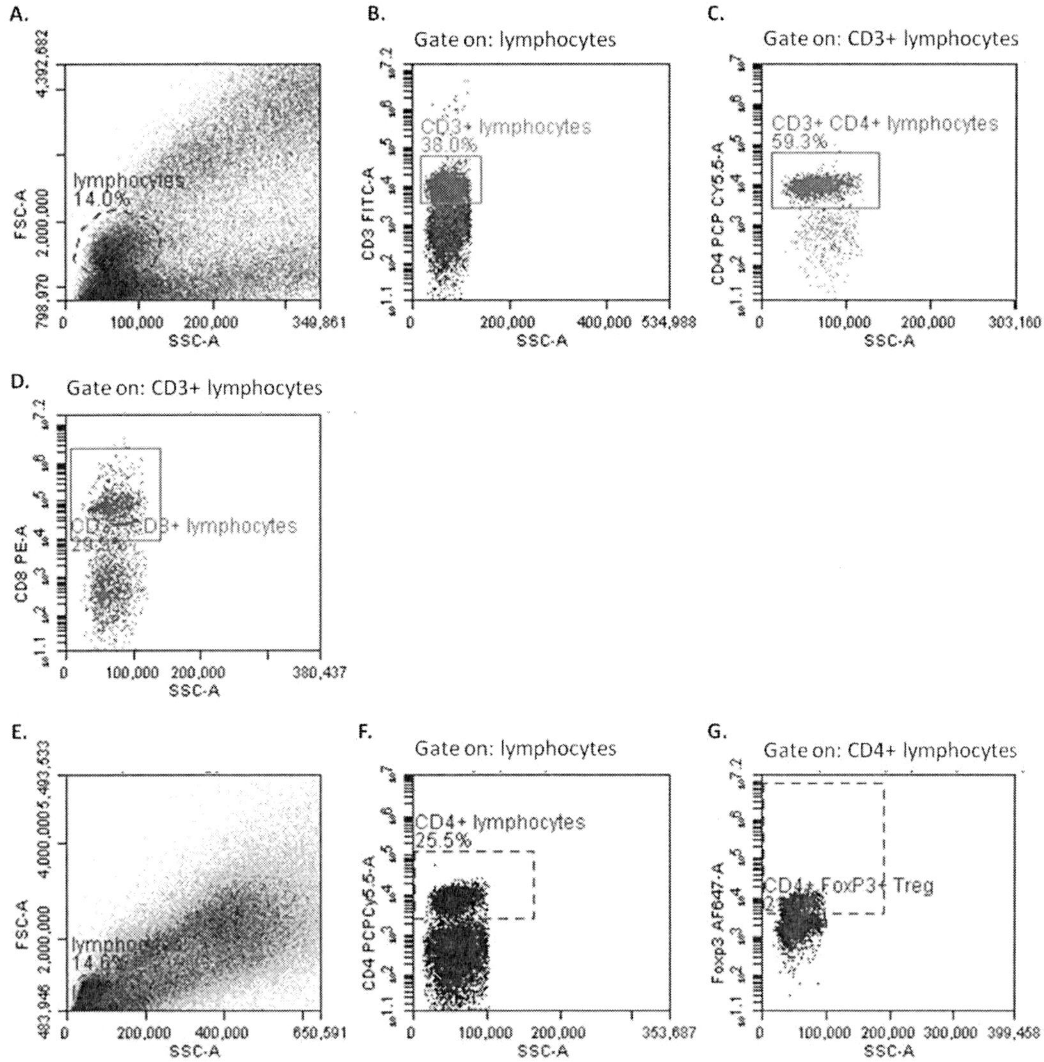

Figure 3. *Gating strategy to identify (regulatory) T cell subsets in mouse colon. A.* Lymphocytes within the LPMC population were identified based on low forward and side scatter values. Dead cells and subcellular debris were eliminated by using a FSC threshold. *B.* Within the lymphocyte gate, T cells were identiefied as CD3^{+} cells. *C.* Within the CD3^{+} T cell population, T helper cells were characterized based on their CD4 expression, and *D.* cytotoxic T cells based on their CD8 expression. *E.* In the second tube, the lymphocytes within the total LPMC were again gated based on their low forward and side scatter. *F.* Within these lymphocytes Thelper cells were identified as CD4^{+}. *G.* Within the CD4^{+} T helper cell population, Treg cells were characterized as CD4^{+} FoxP3^{+}.

In conclusion, Treg cells (characterized as $CD4^+$ $Foxp3^+$ cells) were identified at low frequencies (0.66 ± 0.14%) in the LPMC fraction of healthy control BALB/c mice. However, based on their frequency within the LPMC fraction, they cannot be classified as rare events in the lamina propria of mouse colon. To further characterize Treg subsets in detail, other markers, such as CD25 and CTLA-4, need to be included in the analysis.

Conclusion

The study of rare cell populations is gaining importance and flow cytometry still remains the golden standard for analysis of these rare events. This chapter outlines several key principles that need to be taken into account in order to perform successful analysis of events that appear at low frequency but could be of great importance for the immune response.

Minimization of the background signal, avoidance of autofluorescence using a mock/dump channel, inclusion of multiple parameters - of which at least one negatively marks the rare event – in the analysis, maximization of signal over noise, careful selection of the reagents, correct matching to fluorochromes, proper compensation, running appropriate controls and acquisition of an appropriate number of events will considerably increase your success rate for performing rare event analysis. Sometimes you might need to sacrifice the best option for one parameter to get a better option for another parameter e.g. choose a less bright fluorochrome for a rare event in order to avoid spillover. The examples presented here demonstrate how carefully designed multicolor flowcytometric experiments can yield quantitative data on rare cell populations.

References

[1] Mahnke, YD; Roederer, M. Optimizing a multicolor immunophenotyping assay. *Clin.Lab Med.*, 2007, 27, 469-85, v.

[2] Donnenberg, AD; Donnenberg, VS. Rare-event analysis in flow cytometry. *Clin.Lab Med.*, 2007, 27, 627-52, viii.

[3] Sack, U; Tarnók, A; Rothe, G. *Cellular Diagnostics: Basic Principles, Methods, and Clinical Applications of Flow Cytometry*. 2009

[4] Khan, SS; Solomon, MA; McCoy, JP, Jr. Detection of circulating endothelial cells and endothelial progenitor cells by flow cytometry. *Cytometry B Clin.Cytom.*, 2005, 64, 1-8.

[5] Fischer, JC. *Endothelial Progenitor Cells*. 2009, 305-316.

[6] Monici, M. Cell and tissue autofluorescence research and diagnostic applications. Biotechnol. *Annu.Rev.*, 2005, 11, 227-256.

[7] Hulspas, R; O'Gorman, MR; Wood, BL; Gratama, JW; Sutherland, DR. Considerations for the control of background fluorescence in clinical flow cytometry. *Cytometry B Clin.Cytom.*, 2009, 76, 355-364.

[8] Mayeno, AN; Hamann, KJ; Gleich, GJ. Granule-associated flavin adenine dinucleotide (FAD) is responsible for eosinophil autofluorescence. *J. Leukoc. Biol.*, 1992, 51, 172-175.

[9] Zimmerlin, L; Donnenberg, VS; Donnenberg, AD. Rare event detection and analysis in flow cytometry: bone marrow mesenchymal stem cells, breast cancer stem/progenitor cells in malignant effusions, and pericytes in disaggregated adipose tissue. *Methods Mol.Biol.*, 2011, 699, 251-273.

[10] Donnenberg, VS; Donnenberg, AD. Identification, rare-event detection and analysis of dendritic cell subsets in broncho-alveolar lavage fluid and peripheral blood by flow cytometry. *Front Biosci.*, 2003, 8, s1175-s1180.

[11] Landmesser, U; Spiekermann, S; Preuss, C; Sorrentino, S; Fischer, D; Manes, C; Mueller, M; Drexler, H. Angiotensin II induces endothelial xanthine oxidase activation: role for endothelial dysfunction in patients with coronary disease. Arterioscler. *Thromb.Vasc.Biol.*, 2007, 27, 943-948.

[12] Maecker, HT; Frey, T; Nomura, LE; Trotter, J. Selecting fluorochrome conjugates for maximum sensitivity. *Cytometry A*, 2004, 62, 169-173.

[13] Maecker, HT; Trotter, J. Flow cytometry controls, instrument setup, and the determination of positivity. *Cytometry A*, 2006, 69, 1037-1042.

[14] Baumgarth, N; Roederer, M. A practical approach to multicolor flow cytometry for immunophenotyping. *J. Immunol.Methods*, 2000, 243, 77-97.

[15] Hulspas, R; Dombkowski, D; Preffer, F; Douglas, D; Kildew-Shah, B; Gilbert, J. Flow cytometry and the stability of phycoerythrin-tandem dye conjugates. *Cytometry A*, 2009, 75, 966-972.

[16] Roederer, M; De, RS; Gerstein, R; Anderson, M; Bigos, M; Stovel, R; Nozaki, T; Parks, D; Herzenberg, L; Herzenberg, L. 8 color, 10-parameter flow cytometry to elucidate complex leukocyte heterogeneity. *Cytometry*, 1997, 29, 328-339.

[17] Kraan, J; Gratama, JW; Keeney, M; D'Hautcourt, JL. Setting up and calibration of a flow cytometer for multicolor immunophenotyping. *J. Biol. Regul. Homeost. Agents*, 2003, 17, 223-233.

[18] O'Gorman, MR; Thomas, J. Isotype controls--time to let go? *Cytometry*, 1999, 38, 78-80.

[19] Keeney, M; Gratama, JW; Chin-Yee, IH; Sutherland, DR. Isotype controls in the analysis of lymphocytes and CD34+ stem and progenitor cells by flow cytometry--time to let go! *Cytometry*, 1998, 34, 280-283.

[20] van de Winkel, JG; Capel, PJ. Human IgG Fc receptor heterogeneity: molecular aspects and clinical implications. *Immunol. Today*, 1993 *14*, 215-221.

[21] Roederer, M. Spectral compensation for flow cytometry: visualization artifacts, limitations, and caveats. *Cytometry*, 2001, 45, 194-205.

[22] Martin-Henao, GA; Ingles-Esteve, J; Cancelas, JA; Garcia, J. Isolation of CD34+ hematopoietic progenitor cells in chronic myeloid leukemia by magnetic activated cell sorting (MACS). *Bone Marrow Transplant.*, 1996, 18, 603-609.

[23] Hoy, T. *Rare-event detection*. 2013, 55-58.

[24] Roederer, M. How many events is enough? Are you positive? *Cytometry A*, 2008, 73, 384-385.

[25] Van Vré, EA; Van Brussel, I; Bosmans, JM; Vrints, CJ; Bult, H. Dendritic cells in human atherosclerosis: from circulation to atherosclerotic plaques. *Mediators. Inflamm.*, 2011, 941396-

[26] Van Brussel, I; Bult, H; Martinet, W; De Meyer, GR; Schrijvers, DM. *Dendritic Cells in Atherogenesis: from Immune Shapers to Therapeutic Targets*. 2012.

[27] Granucci, F; Zanoni, I; and Ricciardi-Castagnoli, P. Central role of dendritic cells in the regulation and deregulation of immune responses. *Cellular and Molecular Life Sciences*, 2008, 65, 1683-1697.

[28] Van Brussel, I; Berneman, ZN; Cools, N. Optimizing dendritic cell-based immunotherapy: tackling the complexity of different arms of the immune system. *Mediators.Inflamm.,* 2012, 690643-

[29] Steinman, RM. Dendritic cells: versatile controllers of the immune system. *Nat.Med.*, 2007, 13, 1155-1159.

[30] Van Brussel, I; Van Vré, EA; De Meyer, GRY; Vrints, CJ; Bosmans, JM; Bult, H. Expression of dendritic cell markers CD11c/BDCA-1 and CD123/BDCA-2 in coronary artery disease upon activation in whole blood. *J. Immunol.Methods*, 2010, 362, 168-175.

[31] Donnenberg, VS; O'Connell, PJ; Logar, AJ; Zeevi, A; Thomson, AW; Donnenberg, AD. Rare-event analysis of circulating human dendritic cell subsets and their presumptive mouse counterparts. *Transplantation*, 2001, 72, 1946-1951.

[32] Lindstedt, M; Lundberg, K; Borrebaeck, CA. Gene family clustering identifies functionally associated subsets of human in vivo blood and tonsillar dendritic cells. *J. Immunol.*, 2005, 175, 4839-4846.

[33] Merad, M; Sathe, P; Helft, J; Miller, J; Mortha, A. The dendritic cell lineage: ontogeny and function of dendritic cells and their subsets in the steady state and the inflamed setting. *Annu. Rev. Immunol.*, 2013 31, 563-604.

[34] Ziegler, SF; Buckner, JH. FOXP3 and the regulation of Treg/Th17 differentiation. *Microbes.Infect.*, 2009, 11, 594-598.

[35] Shalev, I; Schmelzle, M; Robson, SC; Levy, G. Making sense of regulatory T cell suppressive function. *Semin.Immunol.*, 2011, 23, 282-292.

[36] Liu, W; Putnam, AL; Xu-Yu, Z; Szot, GL; Lee, MR; Zhu, S; Gottlieb, PA; Kapranov, P; Gingeras, TR; Fazekas de St, GB; Clayberger, C; Soper, DM; Ziegler, SF; Bluestone, JA. CD127 expression inversely correlates with FoxP3 and suppressive function of human CD4+ T reg cells. *J. Exp. Med.*, 2006, 203, 1701-1711.

[37] Mangodt, TC; Van Herck, MA; Ruyssers, NE; Ramet, J; De Dooy, J; De Winter, BY. *Review of the immunological mechanisms and the role of the Th17/Treg balance in the pathogenesis of RSV infection*. In preparation, 2013

[38] Egwuagu, CE. STAT3 in CD4+ T helper cell differentiation and inflammatory diseases. *Cytokine*, 2009, 47, 149-156.

[39] Zhu, J; Yamane, H; Paul, WE. Differentiation of effector CD4 T cell populations (*). *Annu. Rev. Immunol.*, 2010, 28, 445-489.

[40] Setiawan, T; Metwali, A; Blum, AM; Ince, MN; Urban, JF, Jr.; Elliott, DE; Weinstock, JV. Heligmosomoides polygyrus promotes regulatory T-cell cytokine production in the murine normal distal intestine. *Infect. Immun.*, 2007, 75, 4655-4663.

In: Flow Cytometry
Editor: Stefanos Papandreou

ISBN: 978-1-62808-709-3

Chapter 4

Light-Scattering Flow Cytometry: Advanced Characterization of Individual Particle Morphology

Valeri P. Maltsev**, Andrey V. Chernyshev and Dmitry I. Strokotov
Institute of Chemical Kinetics and Combustion, Novosibirsk, Russian Federation
Novosibirsk State University, Novosibirsk, Russian Federation

Abstract

Light scattering has played an important role in flow cytometry since the beginning of its technological development. Most of the significant results were demonstrated with analyses of individual particles from light scattering. However the progress in fluorescent biomarkers reduced the role of light scattering in flow-cytometer analyses substantially. Nevertheless the light scattering is still alive and we are presenting results from a new stage in the development of light-scattering analysis in flow cytometry. These new advances stem from angle-resolved methodology. With this review we show how the angle-resolved light scattering provides determination of physical characteristics of individual particles. Moreover this approach allows one to retrieve particle dimensions with a precision ranging from 10 nm to 100 nm depending on the shape complexity of a particle. This is subdiffraction precision that cannot be realized with conventional optical microscopy. Moreover the flow-cytometer analysis provides statistically proven results for characterization of cell populations because of the high rate of measurements. In order to demonstrate the advantages of angle-resolved measurement of light scattering we describe the principles of scanning flow cytometry that allows us to measure the light-scattering profiles (LSPs) of individual particles. In the development of this approach, we introduce the latest improvement that relates to simultaneous measurement of regular and polarized LSPs for one particle. The methodological goal of the angle-resolved technique is the determination of morphological characteristics of particles from a solution of the

* Valeri P. Maltsev: Phone: +7(383)3333240, E-mail: maltsev@kinetics.nsc.ru, Fax: +7(383) 3307350, Internet: http://cyto.kinetics.nsc.ru/.

inverse light-scattering (ILS) problem. We introduce a few methods to solve the ILS problem utilizing the numerical algorithm in global optimization. The bisphere characteristics, sizes, and refractive indices of each sphere composing the bisphere were successfully retrieved by processing the regular and polarized LSPs. The LSPs showed a near-perfect agreement with T-matrix simulations, resulting in a 50-nm precision for sizing bispheres. We rigorously characterized individual blood microparticles, determining their sizes and refractive indices from measured LSPs. We introduce a novel approach for determination of the volume and shape of individual blood platelets modeled as oblate spheroids from angle-resolved light scattering using the flow-cytometer technique. Accurate measurement of the spheroidal aspect ratio of the platelets allowed us to discover the substantial differences between native and activated 10 μM adenosine diphosphate samples of platelets. We demonstrate a flow-cytometer method to measure the morphology of single *Escherichia coli* cells with a precision ranging from 50 nm to 250 nm and from 5 nm to 25 nm for bacteria length and diameter, respectively. High precision in the determination of cell morphology allows us to measure kinetics of intracellular processes like red-blood-cell lysis and lymphocyte apoptosis. We discuss further applications of angle-resolved light scattering using flow cytometry in the conclusion.

Introduction

Humans encounter many different systems in their habitat that can affect their health: blood, aqua environment, aerosols, dairy products, etc. Quality control of these systems can be realized using the latest achievements in fundamental science to provide continuous improvements of life quality. A disperse system is formed by particles whose physical characteristics are different from surrounding medium. To provide advanced characterization of such disperse systems, we have to support a scientific research by modern technologies in the basic areas of science which are as follows: instrumental, theoretical and experimental. Let us consider the methods that are based on the optic phenomena to analyze particles in a homogeneous medium. These methods can be divided into the following categories: (1) methods based on the analysis of light emitted by a particle suspension; and (2) methods based on the analysis of light emitted by a single particle. Practical methods from the first category can be instrumentally realized in a simple way by shifting the analytical problems into theoretical and experimental areas. Indeed scientific publications demonstrate a complexity of theoretical considerations when light interacts with polydisperse multicomponent systems. Additional processing of experimental data requires substantial simplification of the experimental model to retrieve required particle characteristics forming the disperse system. The methods of the second category are more complex from an instrumental point of view; whereas, theoretical and experimental procedures are considerably simplified for these methods.

The problem of single-particle analysis using optical methods arises in different fields, including astronomy, remote sensing, and analysis of aerosols and emulsions [1]. The methods used in the analysis of *individual* particles, especially biological cells, can be divided into the following categories: (1) probe-field methods forming an electromagnetic field in the smallest volume, like confocal microscopy; and (2) full-field methods based on a detailed analysis of light scattered by a particle. The most powerful method that may be considered to belong to the second category is flow cytometry because the hydrodynamic focusing system provides a means for analysis of individual particles. Flow cytometers allow for an analysis of individual particles in a flow using the following physical phenomenon: fluorescence, light scattering, and conductivity. In fact an overwhelming majority of flow cytometers utilize fluorescence and light scattering to analyze individual particles carrying by a flow. There exist two levels of scientific analysis of particles in a disperse system: identification and characterization. Identification, also known as classification, refers to attributing the particles to one of several, usually predefined, classes from experimentally measured signals. Characterization refers to the quantitative measurement of a particle by its morphological and functional characteristics, which can be retrieved from a solution of the inverse problem. Evidently, characterization requires substantial scientific research to establish a mathematical relation between experimental signals and characteristics of a particle. An overwhelming majority of flow-cytometer studies solve just identification problem in the analysis of disperse systems by counting and sorting particles based on their measured intensities of fluorescence and light scattering. In the analyses, the intensity of fluorescence depends on the amount of dye molecules that are linked to the particular molecular components specifically. This method is very useful to identify biological cells with different membrane or/and intracellular proteins. In rare instances, the fluorescence is utilized to solve the characterization problem of cell analysis with the determination of the distribution of specified membrane proteins over a cell population [2, 3]. Indeed the determination of the distributions required development of an adequate mathematical model of ligand-receptor-binding kinetics and solution of the inverse problem using fluorescence measurement and global optimization [4]. This approach allows the determination of a rate constant of the ligand-receptor binding that relates to the characterization of cell population.

In this chapter we would like to clarify the current state of the art of analysis of individual particles from light scattering with flow cytometry by focusing on the solution of the characterization problem. What do we mean by the solution of characterization problem from light scattering? In comparison to fluorescence the light scattering is much complex phenomena from instrumental, theoretical and experimental points of view. Characteristics of scattered light depend on the physical characteristics of a particle, orientation of the particle relative to direction of illumination, and on the characteristics of incident light that include its intensity, wavelength, and polarization. The light scattered in a fixed angle is characterized by an intensity and polarization assuming the same wavelength as for incident light. The dependence of scattered-light characteristics on angle forms the spatial distribution of light scattering, or the light-scattering pattern. A particle is characterized by a volume, shape, and internal structure, that is characterized through its refractive index distribution. The orientation of the particle relative to the direction and polarization of incident light also must be taken into account. So the solution of the characterization problem from light scattering means determination of particle characteristics and orientation from the measurement of the properties of the incident light and light-scattering pattern. In order to solve the characterization problem one should keep the following trajectory: light-scattering measurement --> particle identification --> optical model of the particle --> theoretical simulation of light scattering --> solution of the inverse light-scattering problem. The first and second stages are determined by an instrument; the third and fourth stages are determined by a theory; and the last stage is determined by experimental procedures, i.e. processing of experimental data. There are not so many studies published in articles where this trajectory has been proposed. Wyatt passed this way for the first time with the Differential I and II as instruments, with the Mie theory and Rayleigh-Gans approximation as theories, and the least-square fit as an experimental procedure [5].

In actuality, the relationship between particle characteristics and light scattering are rather complex and most of them are not available in closed analytical form. Hence any description of these relations assumes a usage of significant of mathematical definitions and expressions, calculating formulas and algorithmic codes. With this chapter we try to introduce these relations between particle characteristics and light scattering in a verbal form with references to original publications where mathematics has been introduced in detail.

Instrument

Light-scattering Capabilities of Ordinary Flow Cytometers

An ordinary flow cytometer allows one to measure only two numbers from light scattering. These are the intensities of forward scatter (FSC) and side scatter (SSC). By plotting these values on a two-dimensional map, one can solve the identification problem successfully, for example, for lymphocytes, monocytes and granulocytes of blood. However, the relationship between cell morphology and these two values is generally so complicated that little hope is left for a detailed characterization of disperse system. However, ordinary flow cytometry with small modifications can allow one to solve the identification problem for blood granulocytes [6]. The modification assumes the measurement of the SSC signals with and without a polarizer and plotting these intensities on a two-dimensional map to count eosinophils and neutrophils separately. This instrumental achievement was proven theoretically with what is currently the most powerful light-scattering theory, the discrete dipole approximation, which allows the simulation of light scattering of a particle with an irregular shape and internal structure [7]. Unfortunately today's experimental procedures do not provide a retrieval of characteristics of granulocytes from light scattering because of the complexity of the inner structure of these cells. In addition, there is one example where the ordinary flow cytometer characterizes an individual particle from light scattering. In order to characterize red blood cells (RBCs) with the flow cytometer as an instrument, Tycko et al. measured the light scattered at two forward scattering angles [8]. The Mie theory was applied to simulate the light scattering from RBCs, varying volume and hemoglobin concentration. The Mie theory allows the simulation of light scattering from homogeneous spherical particles, but a mature RBC has a biconcave disk shape. In order to account for this discrepancy, the original biochemical treatment was applied to spherize RBCs with a constant volume. The solution of the inverse light-scattering problem was realized by the precomputed mesh database with the linear interpolation between meshes and with a calibration by particles with known volume and refractive index. This approach was modernized with the scanning flow cytometry where calibration was eliminated from the data processing procedure [9].

Angle-resolved Light Scattering for Identification and Characterization

The main problem of ordinary flow cytometry in characterization of individual particles from light scattering is insufficient experimental data. The FSC and SSC signals allow for the determination of particle characteristics modeled by just a homogeneous sphere in fixed bounds for size and refractive index. A logical development of ordinary flow cytometers is to measure the light scattering at multiple angles or even the angle-resolved light-scattering profile (LSP) of individual particles over a wide angular range. Unfortunately, there are only a very small number of studies where multiangle light scattering has been used in the solution of identification and characterization problems in analysis of disperse systems. We completely agree with the statement by Rajwa et al., "Owing to the immense progress in fluorescent label development, multiparameter flow cytometry moved in the last decade in the direction of adding more fluorescence detectors, rather than enhancing scatter measurements" [10]. Nevertheless angle-resolved light scattering is still alive and we anticipate that it will have a significant impact on flow cytometry, especially on the morphological characterization of biological systems.

Historically, angular-resolved flow cytometry was used initially in the solution of the identification problem either with a multiangle instrument [11] or with scanning systems in orthogonal [12] and parallel [13] configurations between illumination and flow. The angle-resolved approach can be realized in one dimensional (1D) LSPs, polar scattering angle, or two dimensional (2D) LSPs, polar and azimuthal scattering angle. Obviously 2D LSP instrument is more complex in measurement of light scattering from individual particles comparing to 1D LSP instrument. As usual recording of 2D LSP requires a high-speed matrix detector with high-photon sensitivity. Nevertheless, the 2D LSP instruments were demonstrated as the next version of the scanning flow cytometer [14], diffraction imaging flow cytometer [15, 16], and interferometric two-dimensional scattering system [17]. At present the angle-resolved light scattering still has not realized the identification potential in an analysis of disperse systems passing the lag-phase in development [18].

Let us move to consider solutions of the characterization problem with flow cytometry in the analysis of individual particles forming disperse systems. The first attempt to solve this problem was demonstrated by Bartholdi et al. [19] with a custom-made flow cytometer equipped with a rather complex multiangle detecting system. They used the Mie theory to simulate light scattering from latex spheres. Processing of experimental data was performed with a point-by point comparison of simulated and measured intensities at different scattering angles. A few samples of latex spheres were successfully characterized from their sizes, fixing the refractive index of polysterene. Unfortunately the authors did not develop a mathematical procedure to estimate the precision of sizing of spheres from the characterization method.

The next stage of development of angle-resolved light scattering and solution of the characterization problem began with the introduction of the scanning-flow cytometer (SFC) [20]. We utilized the scanning principle of Loken et al. [12], but coincided the directions of flow and illumination of a particle. Instrumentally the SFC allows for the measurement of the entire angle-resolved LSP of an individual particle in the angular range from 5 to 100 degrees using one detector and spherical mirror as a scanning system. It looks very simple compared to other experimental set-ups for measuring the angle-resolved light scattering. At present we are using the second generation of the SFC fabricated by CytoNova company (Novosibirsk,

Russia, http://cyto.kinetics.nsc.ru/) as a commercial prototype (Figure 1). The current set-up of the SFC allows us to measure three fluorescence signals in different spectral regions and two light-scattering profiles, regular and polarized, of individual particles with a rate of 300 particles per second. The SFC is equipped with three lasers with wavelengths: 405 nm and 660 nm for scattering and 488 nm for fluorescence. The third generation of the SFC is coming soon.

With the instrumental problem of the measurement of angle-resolved light scattering effectively solved, our activity focused on the development of the computational algorithms for simulation of light scattering using the latest theoretical achievements. These algorithms have to simulate light scattering of a particle with an irregular shape and an internal structure. There was just one general approach in light scattering theories – discrete dipole approximation (DDA). The main problem of the DDA simulation is the requirement of a huge amount of computer memory and high computer clock rate. Fighting this problem, we developed the effective code ADDA that allows for the simulation of light scattering from rather large particles with reasonable computing times [21, 22]. At present the ADDA allows us to simulate light scattering of the most complex blood cell, like a neutrophil [23].

Figure 1. The commercial prototype of the scanning flow cytometer.

The final stage in the development of technology to characterize disperse systems from angle-resolved light scattering is support of this technology through effective processing of experimental data. The complex shape and inner structure of an analyzed particle requires a complex optical model of this particle with a number of physical characteristics of the particle. The homogeneous sphere characterized by size and refractive index is the simplest model that relates to polymer beads, blood microparticles, etc. Two concentric spheres which can model a mononuclear cell already have four characteristics, two diameters and two refractive indices. A dimer formed by two spheres can be characterized by two diameters and two refractive indices but orientation of the dimer relative to direction of the incident light and polarization adds two parameters that must be retrieved from experimental data. A least-square surface formed by simulated and measured data for a particle with more than two characteristics has a substantial number of local minima. Hence, the processing of experimental data has to be done with global optimization to find the global minimum on the surface. We found that DIRECT is the most effective algorithm for global optimization [24]. We supported this algorithm with the procedure to estimate errors of the evaluated characteristics [25].

At present the solution of the characterization problem for individual particles from light scattering is supported by modern implementations in instrumental, theoretical, and experimental areas keeping in mind the SFC, the ADDA, and the DIRECT respectively. Using this approach we solved the characterization problem for a few important particles that are as follows: mononuclear cells [21], blood microparticles [26], blood platelets [27], and rod-like bacterial cells [28]. The maximal result was achieved in the solution of the characterization problem for a dimer of spheres with determination of four characteristics of the dimer and two angles of the dimer orientation [29]. Six characteristics of the inverse light-scattering (ILS) problem can only be realized with measurement of polarized light scattering.

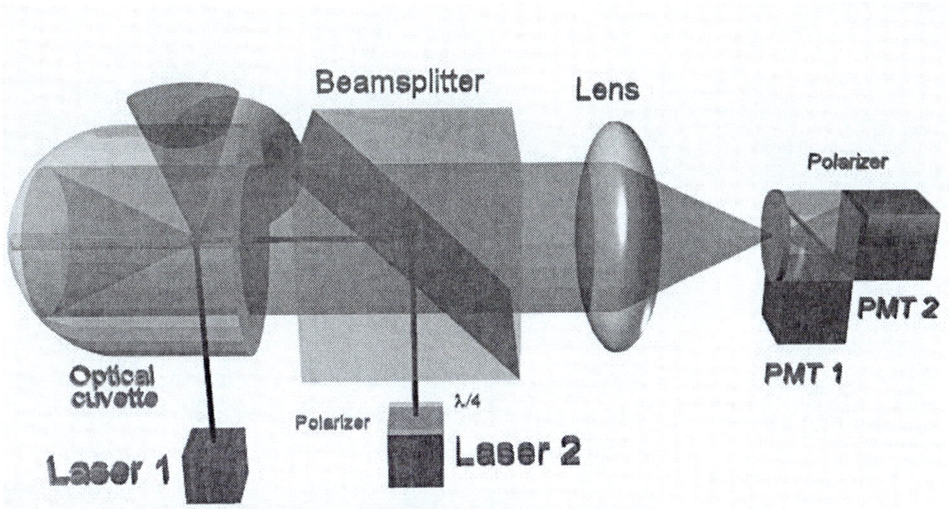

Figure 2. The schematic layout of the optical part of the scanning flow cytometer.

Polarized Light-scattering Flow Cytometry

The current set-up of the SFC allows us to measure the regular and polarized LSPs of individual particles [25]. A detailed description of the measuring principle of the SFC can be found elsewhere [30]. The schematic layout of the SFC optics is shown in Figure 2. The left part of the scheme corresponds to an ordinary flow cytometer. The sheath and probe flow hits the capillary from the left edge of the Optical cuvette. When the particle crosses the beam from Laser 1, it emits the FSC and SSC as well as side fluorescence (shown as cones in Figure 2). In addition to the conventional orthogonal illumination, the beam from Laser 2 is directed opposite to the direction of the flow. This beam illuminates the particle with a constant intensity when the particle is moved from the left to the right edge of the Optical cuvette in the capillary. Light (shown as a cone, a cylinder, and again a cone in Figure 2 sequentially) scattered by the particle is reflected by the spherical mirror that is to the left side of the Optical cuvette and is collected by the Lens into diaphragm of the detection unit. A spherical mirror is not a perfect focuser of light because of the aberration of the spherical surface. As a result, there is just one angle for any particle location on the capillary axis at which the scattered light will be reflected in a direction parallel to the flow and hence focused by the lens to detector unit. Schematically this function is shown in Figure 3 (left).

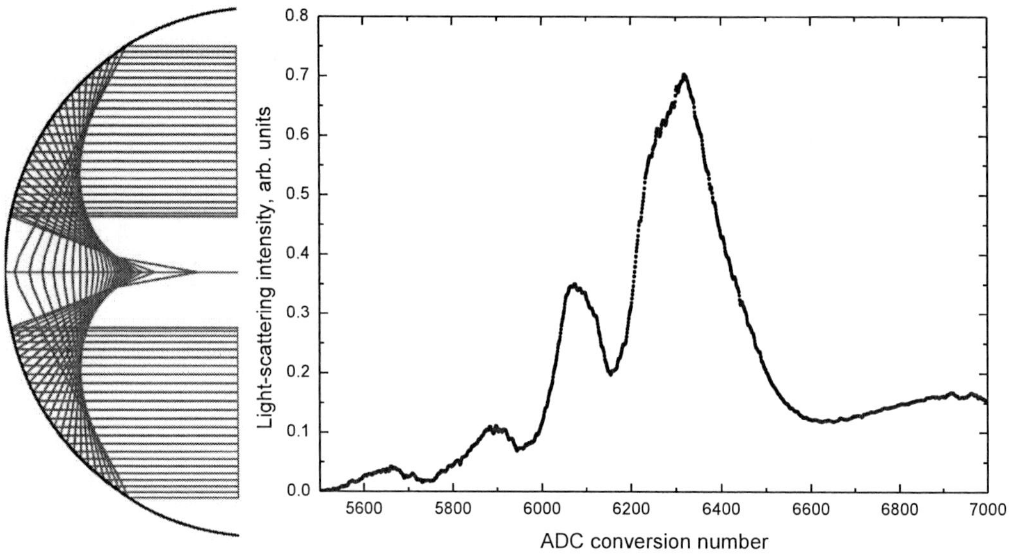

Figure 3. The SFC principle (left). The light-scattering trace of single polymer sphere (right).

Consequently when a particle is moved from the left to the right sides of the Optical cuvette (Figure 2), the detector unit measures the light scattered by this particle from 120 to 5 degrees continuously, resulting in the entire light-scattering trace shown in Figure 3 (right). Using the geometry of the optical unit of the SFC the trace can be transformed to the angle-resolved LSP [26]. The detection unit consists of two channels for measurement of regular and polarized LSPs, PMT 1 and PMT 2 with a Polarizer, respectively (Figure 2). The current from the PMTs is processed by an analog-to-digital converter (ADC). We emphasize important features of the SFC that are the integration of the scattered light over an azimuthal angle and circular polarization of the incident light of the Laser 2 formed by the Polarizer and $\lambda/4$ plate (Figure 2). From these features the regular LSP for a nonspherical particle is independent of the azimuthal angle of the particle orientation that reduces characteristics that have to be retrieved from the solution of the ILS problem by one. Finally the current setup of the SFC supports the solution of the characterization problem with the signals shown in Figure 4. Whereas an ordinary flow cytometer gives just two numbers, for example, the amplitudes of FSC and SSC, the SFC allows one to use two signatures of a particle, regular and polarized LSPs (Figure 4).

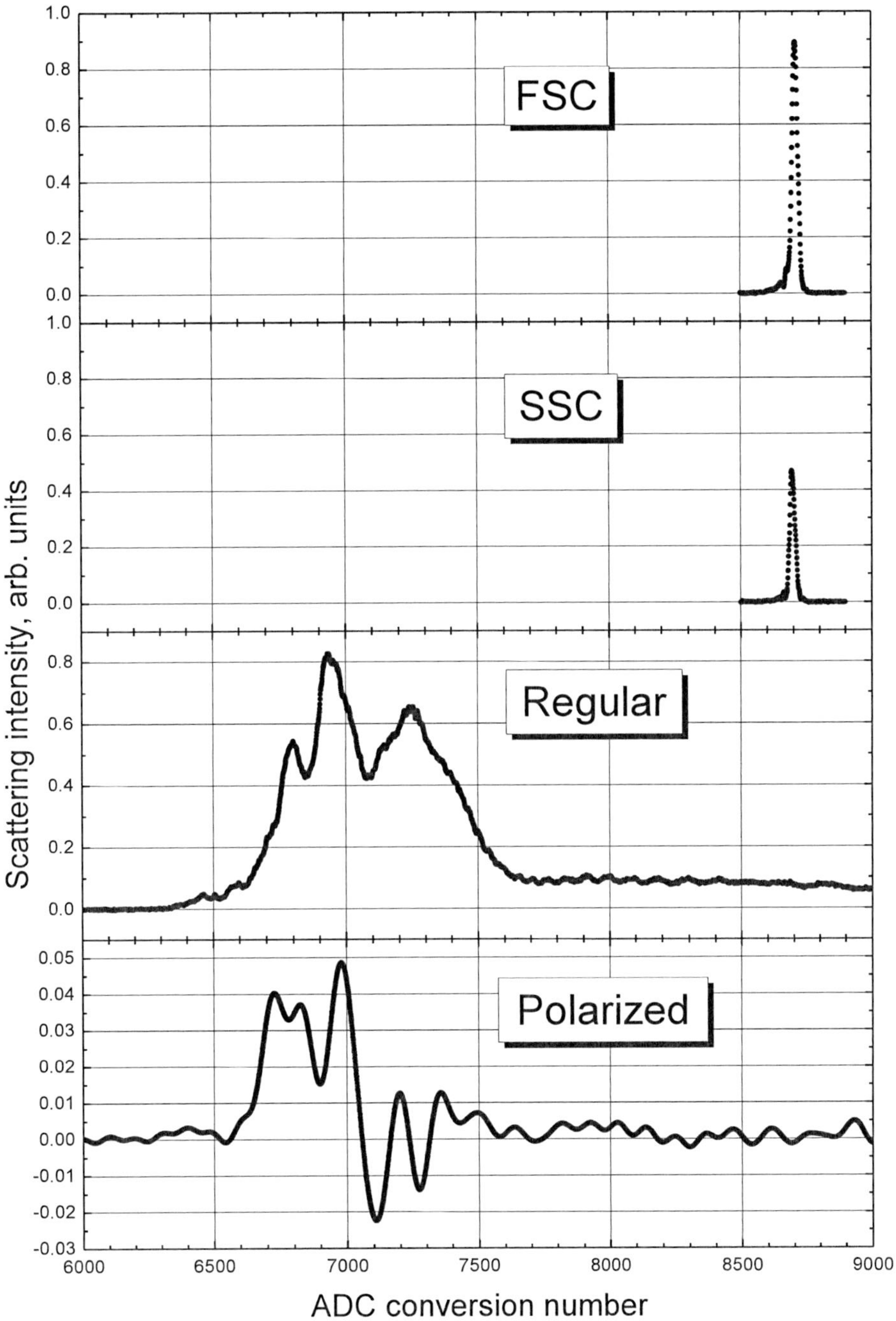

Figure 4. The measured signals of the SFC.

In order to follow the trajectory mentioned above for characterization of particles we have to develop an appropriate optical model of the particle analyzed. A more complex particle in shape and internal structure requires the development of a more complex optical model with a sufficient number of model parameters. A more complex particle model requires more independent data from light-scattering measurement. The measurement of polarized scattered light may provide up to 7 independent LSPs that will be incorporated into future light-scattering instruments. This activity is extremely important for hematology where a solution of the ILS problem could be used for more precise characterization of blood cells. The precise characterization of blood cells assumes measurements of not only concentrations that are inherent for modern hematological analyzers, but also morphological characteristics such as volume, shape, density, nucleus size, etc.

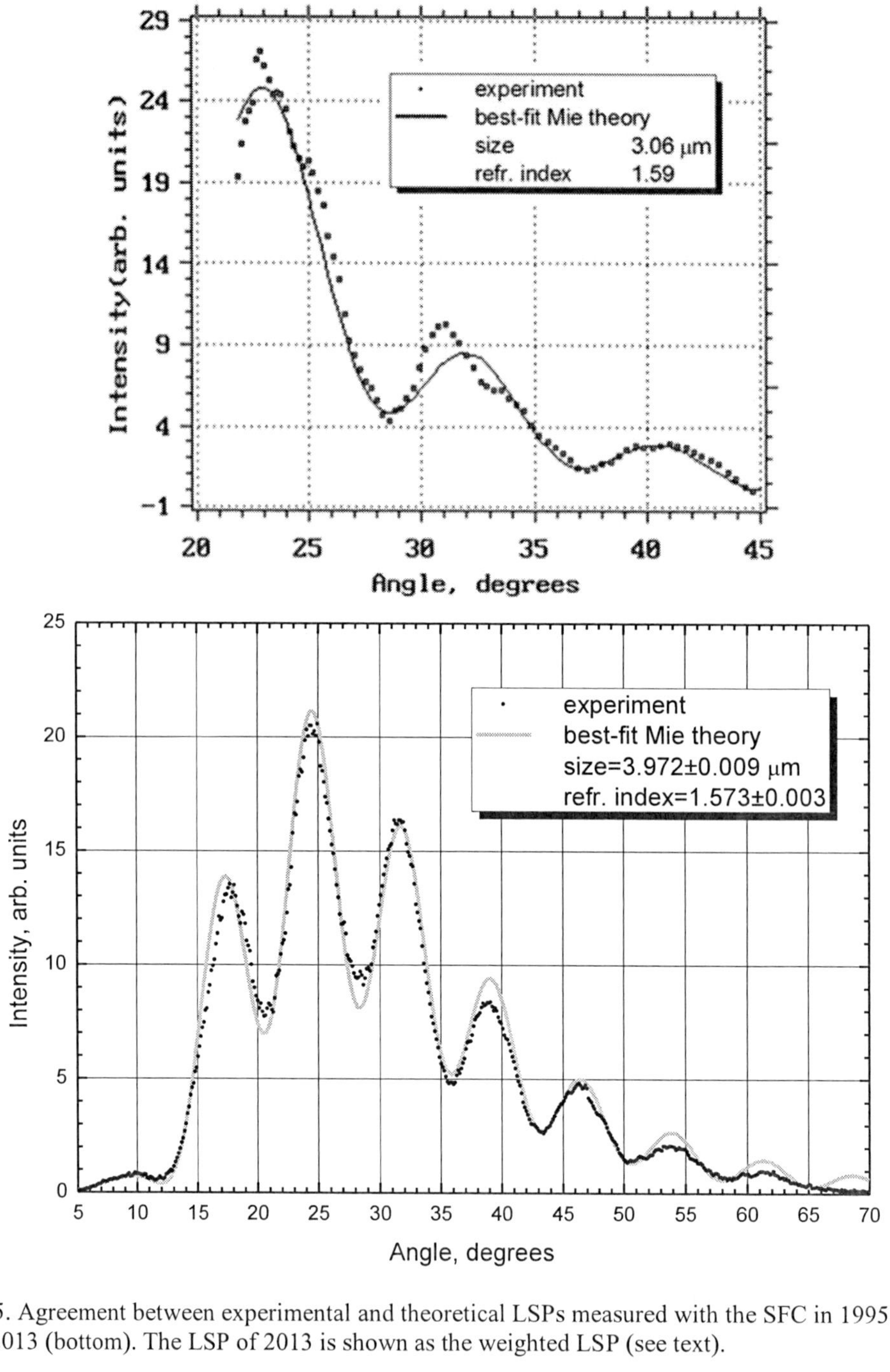

Figure 5. Agreement between experimental and theoretical LSPs measured with the SFC in 1995 (top) and in 2013 (bottom). The LSP of 2013 is shown as the weighted LSP (see text).

Application

Spherical Particles

Monomer

Polymer beads are the most widely available spherical particles for routine analysis with light scattering and flow cytometry. Obviously, the characterization problem for spherical particles was solved by analyzing light scattering from polymer beads. The instrumental implementation of the solution was realized with ordinary flow cytometry measuring light scattering at two spatial angles and producing the two-dimensional map from the signals measured. The map was calibrated from measurement of beads with known mean size and refractive index using the Mie theory as a theoretical implementation. The processing of the experimental data was performed with an interpolation of the measured and calibrated map points.

Angle-resolved light scattering also was utilized to characterize polymer beads. Using scanning flow cytometry, the LSPs of polymer beads were introduced in first publication in 1995 [16]. The SFC demonstrated rather good agreement between the experiment and the Mie theory (Figure 5 top) but the latest experiments with polymer beads showed the near-perfect agreement that results in nanometer precision for spherical particle sizing (Figure 5 bottom). At the moment the operational angular range of the SFC optics and electronics covers the region from 5 to 70 degrees. The operation angular range is substantially expanded with the current set-up of the SFC that forces us to multiply the LSP by the weighted function [25]. The resulting weighted LSP (wLSP) can be represented with a linear scale for the light scattering intensity (Figure 5 bottom).

Dimer

Characterization of a particle from light scattering assumes a solution of the ILS problem, i.e., determination of particle characteristics from light-scattering data. In general, the precision of solutions of the ILS problem depends on the amount of independent light-scattering data and the characteristics of the particle. The regular LSP can be used successfully in the solution of the ILS problem if the number of particle characteristics is increased from two to four. If the amount of particle characteristics exceeds four, the accuracy of the solution of the ILS problem drops dramatically. More independent light-scattering data must be measured to solve the ILS problem for a particle with six characteristics, like a bisphere. The solution of the ILS problem for a bisphere assumes determination of the following characteristics: radii and refractive indices of two spheres composing the dimer and two Eulerian angles. The Eulerian angles describe the orientation of a particle relative to the direction of the incident laser beam and default polarization of optical system. We solved the ILS problem for a bisphere using the SFC as an instrument, the T-matrix method [31] as a theory to simulate light scattering from a bisphere, and the global optimization with DIRECT algorithm as a procedure to process experimental data. The detailed mathematical description of analysis of dimers from regular and polarized LSPs could be found in our publication [25].

The regular and polarized wLSPs were together merged (Figure 6), and the DIRECT algorithm was applied to the joint wLSP by varying six characteristics. Finally, the solution of the ILS problem for a single bisphere allowed us to find the best-fit T-matrix LSP to determine the four characteristics of the bisphere and the Eulerian angles. The result of the solution of the ILS problem for the bisphere is shown in Figure 6. We show the best-fit T-matrix LSP, the characteristics of the bispheres, and the Eulerian angles. The errors of determination of the bisphere characteristics and the Eulerian angles also are presented.

In our studies, we analyzed 500 dimers. The dimer spheres were sized with a mean error of 49 nm (standard deviation of 11 nm) that is extremely precise for optical methods with the wavelength from the visible region. The refractive index of dimer spheres was determined with a mean error of 0.013 (standard deviation of 0.005). The results of the measured distribution of the Eulerian angles are not unexpected: the orientation of the bisphere with angle α is equiprobable over the varying range of the angle; whereas, the orientation of bispheres with angle β is rather narrow, and the mean angle β lies close to 15 degrees. The orientation at an angle of 15^0 corresponds to a rotation of nonspherical particles in Poiseuille flow [26] within a flow channel of the SFC.

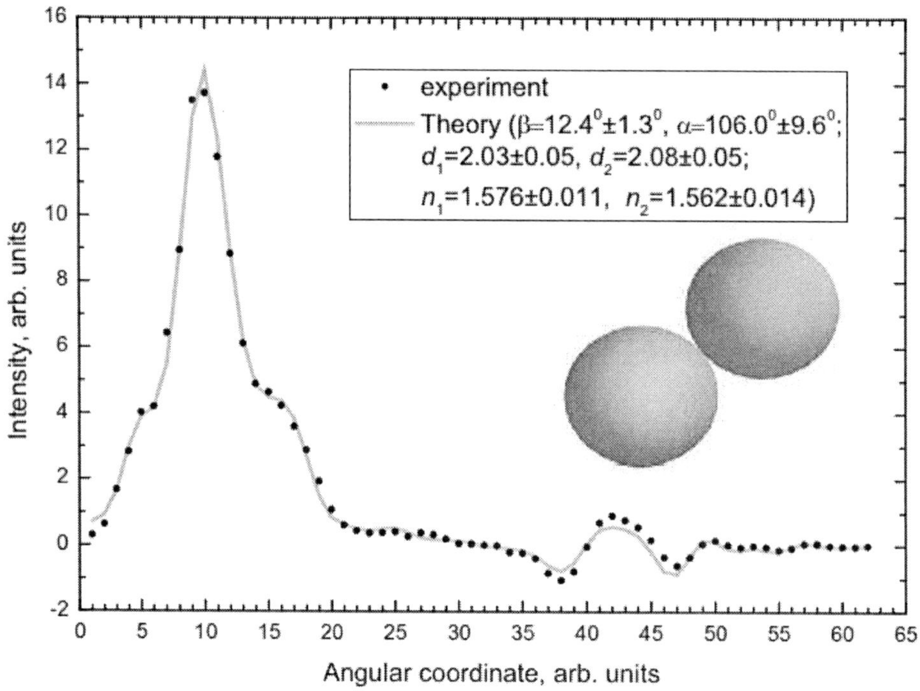

Figure 6. The solution of the inverse light-scattering problem for a bisphere. The points are the experimental merged regular and polarized wLSP. The line is the theoretical best-fit wLSP from T-matrix method.

Rod-like Bacterial Cells

The dimer of spheres could be exactly modeled by two spheres in contact. At present this is the most complex exact optical model of particle which can be characterized from angle-resolved light scattering. There is another category of biological particles that have rather complex shape and internal structure but can be reasonably modeled be a simple physical body. For instance mature blood platelets and *E. coli* cells can be modeled by an oblate spheroid and capped cylinder respectively. The discrepancy between a real biological particle and its optical model results in a systematic error in characterization of the particle from the solution of the ILS problem. Nevertheless, the characterization of a biological population of cells from light scattering has given very reasonable results that are comparable with a microscopic analysis of the populations [24].

We used angle-resolved light scattering to characterize individual *E. coli* cells. The SFC was used to measure the LSPs of cells, the ADDA code was used to simulate of light scattering from capped cylinder and the nearest-neighbor interpolation of a preliminarily calculated database of theoretical LSPs to process experimental data. We were not able to use the global optimization with the DIRECT algorithm because of the iterations of this method. The iteration requires multiple calculations of the direct light-scattering problem with the ADDA that require significant computations. We provide one example that shows the result of processing one wLSP of the *E. coli* cell in Figure 7. This result demonstrates better agreement between the experiment and capped cylinder in comparison to the sphere. Again the angle-resolved light scattering results in near-perfect precision for measurement of length and diameter of rod-like bacterial cells. This means that this method can be used in detailed measurement of elongation of cells during the cell cycle.

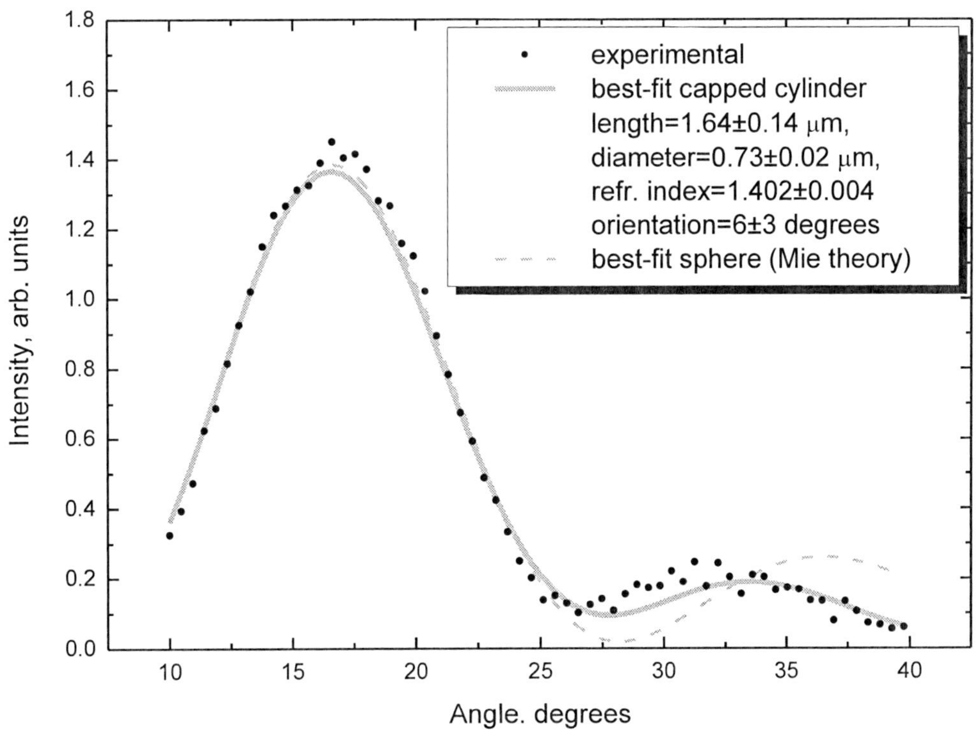

Figure 7. The experimental wLSPs of *E. coli* cell (points) and theoretical best-fit capped cylinder (solid line) and sphere (dashed line).

The method was applied to two strains of *E. coli* cells, showing 135 and 15 nm median precision in determination of length and diameter of single cells, respectively, which is very good for optical methods. We also compared population distributions over model parameters with optical microscope measurements and obtained good agreement for both diameter and length [24].

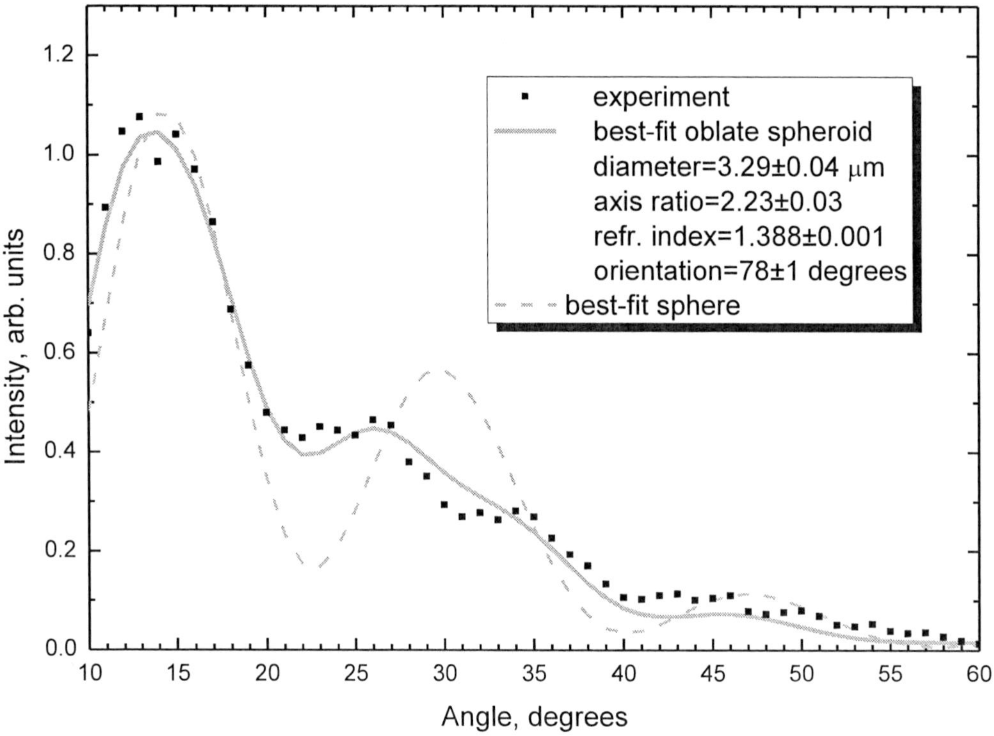

Figure 8. The experimental wLSP (points) of a blood platelet and theoretical best-fit wLSPs of oblate spheroid (line) and sphere (dashed line).

Blood Platelets

Blood platelets have an irregular shape and internal structure but the dimensional size of platelets is close to the wavelength of visible region. Hence we are able to eliminate the small irregularities of cell membrane and internal organelles from the optical model of the cell. Using scanning electron microscopy, a platelet can be modeled by an oblate spheroid. In this way we have to remember that such optical modeling results in a systematic error in the solution of the ILS problem for a mature blood platelet. Similar to our work with *E. coli* cells, we developed the approach to characterize blood platelets by measuring the LSPs by the SFC (instrument), simulating light scattering from the DDA (theory), and performing nearest-neighbor interpolation within the LSP database (data processing) [23]. We apply this approach to blood platelets to characterize both resting and activated cells. The result of the solution of the ILS problem for one platelet is shown in Figure 8. There is rather good agreement between experiment and best-fit theoretical wLSPs; whereas, the best-fit sphere does not follow the experimental points over the angular interval measured with the SFC. Again we emphasize the precision of the determination of the diameter and the aspect ratio of the platelets. High precision is important in measurements of processes with platelets, for example platelet activation and aggregation. The high precision guarantees high sensitivity of external processes acting on blood platelets. In particular we were able to measure a change of the mean aspect ratio for platelets after activation by adenosine diphosphate [23].

This method of characterization of blood platelets was used in the analysis of a donor's blood sample with the measured distribution over the platelet volume. The corresponding distributions over the platelet volume were in a qualitative agreement with the conventional curves measured by Coulter counters. The mean platelet volume values of 10.0 fl for mature platelets fall within the reference range of 8 to 12 fl [23].

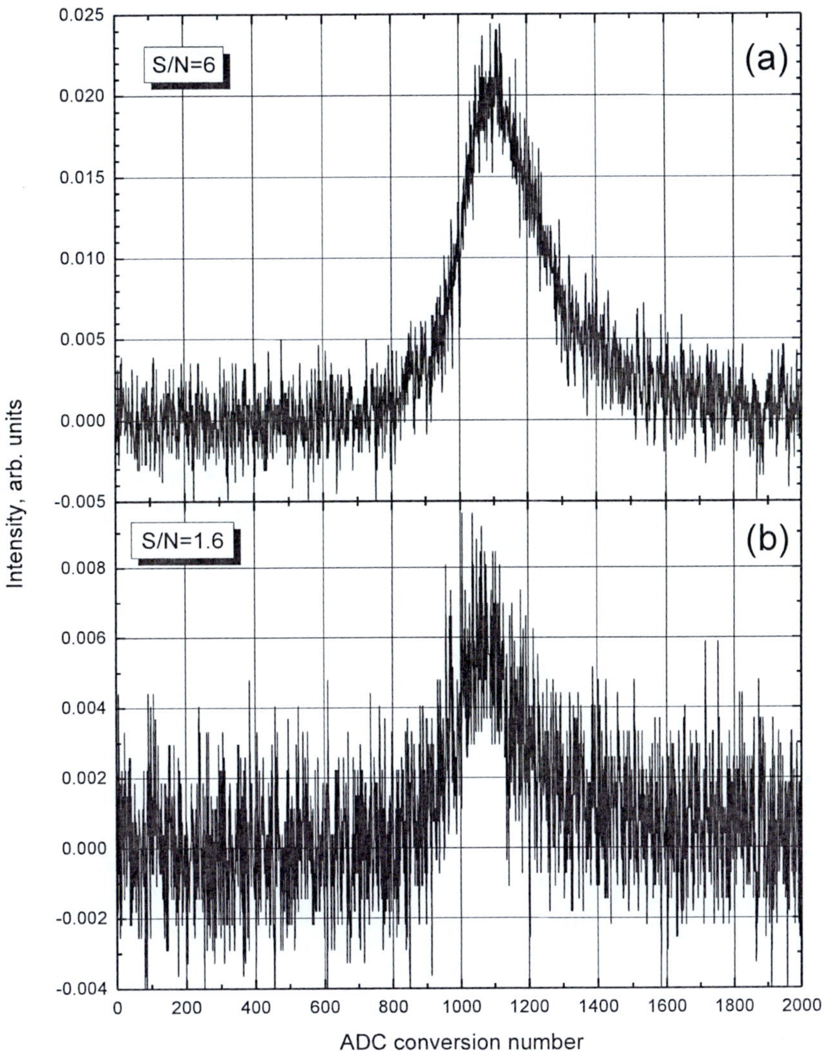

Figure 9. Light-scattering traces of blood microparticles.

Blood Microparticles

The main problem in the measurement of angle-resolved light scattering of blood microparticles is the extremely small scattering intensity because the amplitude of the LSP rapidly decreases with size, resulting in a natural detection limit for any particle. On the other hand, microparticles can provide an instrumental limit for individual particle characterization from light scattering. From scanning electron microscopy images, we can see that the sphere is an appropriate model for a blood microparticle. Consequently, the characterization problem can be solved with the SFC (instrument), the Mie theory (theory), and global optimization (experimental data processing).

We measured a few thousands of microparticles with the SFC. We did not expect oscillations in the LSPs because of the small size of the microparticles. In order to increase the signal-to-noise ratio (S/N), we reduced the angular resolution of the SFC by increasing the diaphragm of the detection unit (Figure 2). We note that the integration of the light scattered by a particle over azimuthal angle by the spherical mirror of the SFC helps in the minimization of the detection limit for small microparticles. We used a laser with a wavelength of 405 nm for the same also to increase the lower detection limit. There are two light-scattering traces from the detecting unit of the SFC in Figure 9. The measurement demonstrates S/N ratios of 6 and 1.6 for Figure 9 (a) and Figure 9 (b) traces respectively. These traces were transformed into the LSPs and then the microparticle characteristics were retrieved using the global optimization with DIRECT algorithm. The characterization gives the following characteristics of the micrioparticles: a size of 0.624±0.010 μm and a refractive index of 1.394±0.003 for the microparticle shown in Figure 9 (a); a size of 0.324±0.008 μm and a refractive index of 1.450±0.006 for the microparticle shown in Figure 9 (b). These experiments have shown that the current set-up of the SFC has a detection limit of 300 nm for biological particles. The detection limit for polystyrene beads can be estimated as 230 nm because of the significantly higher refractive index of 1.58 for polystyrene. This will be established in coming experiments with the SFC.

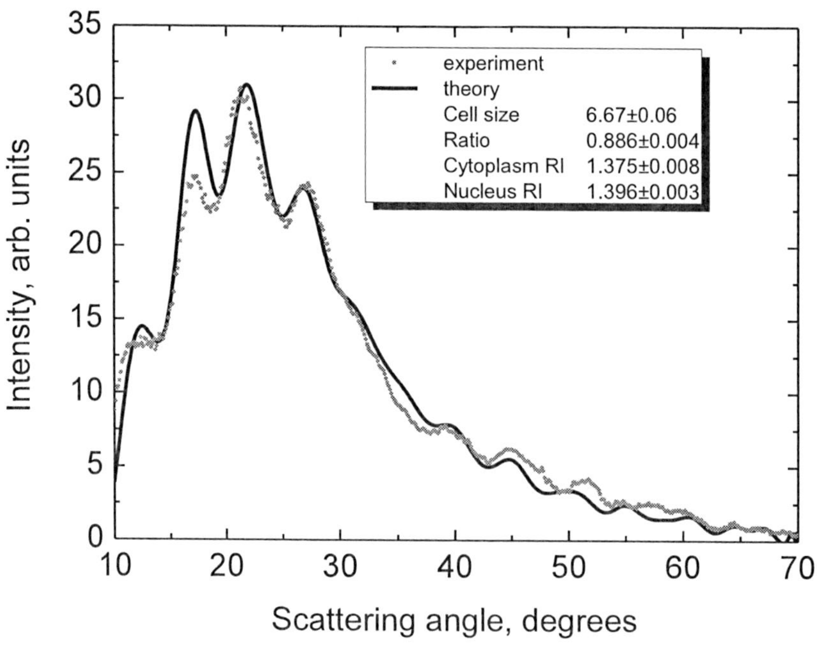

Figure 10. The experimental (points) and theoretical best-fit (line) wLSPs of lymphocyte.

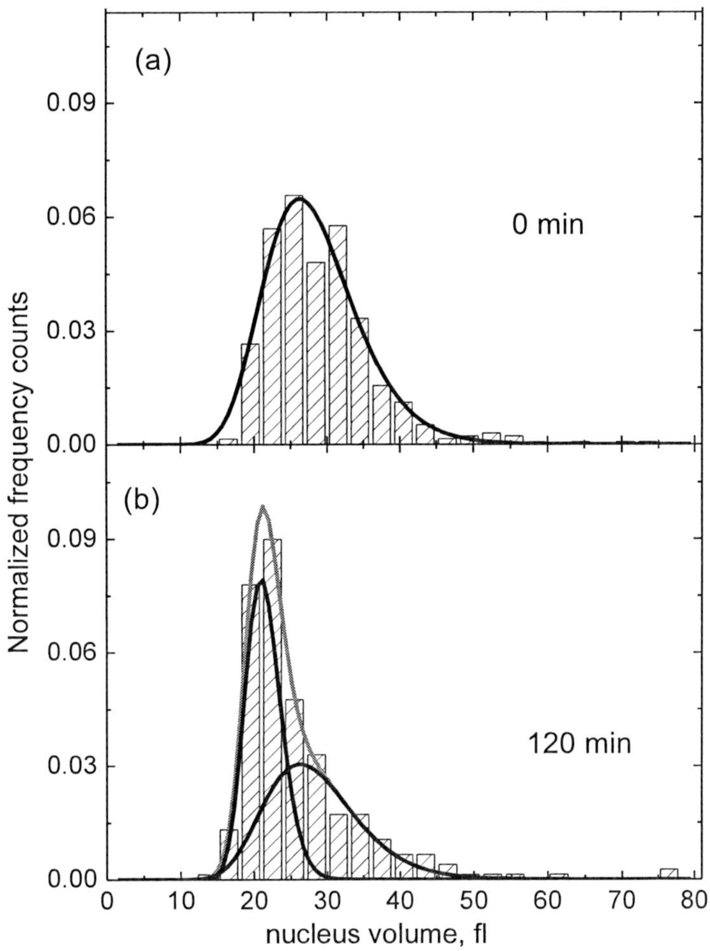

Figure 11. Distributions of lymphocyte nucleus volume (a) initial and (b) in apoptosis.

Mononuclear Cells

At present one more category of biological particles, mononuclear cells, can be characterized from light scattering with flow cytometry. The sphere with a cover is the simplest optical model for a mononuclear cell. Obviously this model is in very rough agreement with the membrane shape and nuclear structure of a real cell, and we expect this to result in a considerable contribution of the systematic error to cellular characteristics retrieved from a solution of the ILS problem. We have shown that one primary source of systematic error arises from the inhomogeneity of the cell nucleus [21, 32]. Hence the errors of characteristics estimated from the solution of the ILS problem could be an indicator of the degree of inhomogeneity of a cell nucleus.

The characterization problem for a mononuclear cell was solved with the use of the SFC (instrument), the Mie theory (theory) and global optimization with DIRECT algorithm (processing of experimental data) [21]. The LSPs of human lymphocytes were measured with the SFC. The optical model for a mononuclear cell has four characteristics, which are as follows: the cell size, the ratio of the nucleus size to the cell size, the cytoplasm refractive index and nucleus refractive index. An example of the characterization of one lymphocyte is shown in Figure 10. There is good agreement between the experimental points and the best-fit Mie theory curve that provides sub-diffraction-precision measurements of the cell size.

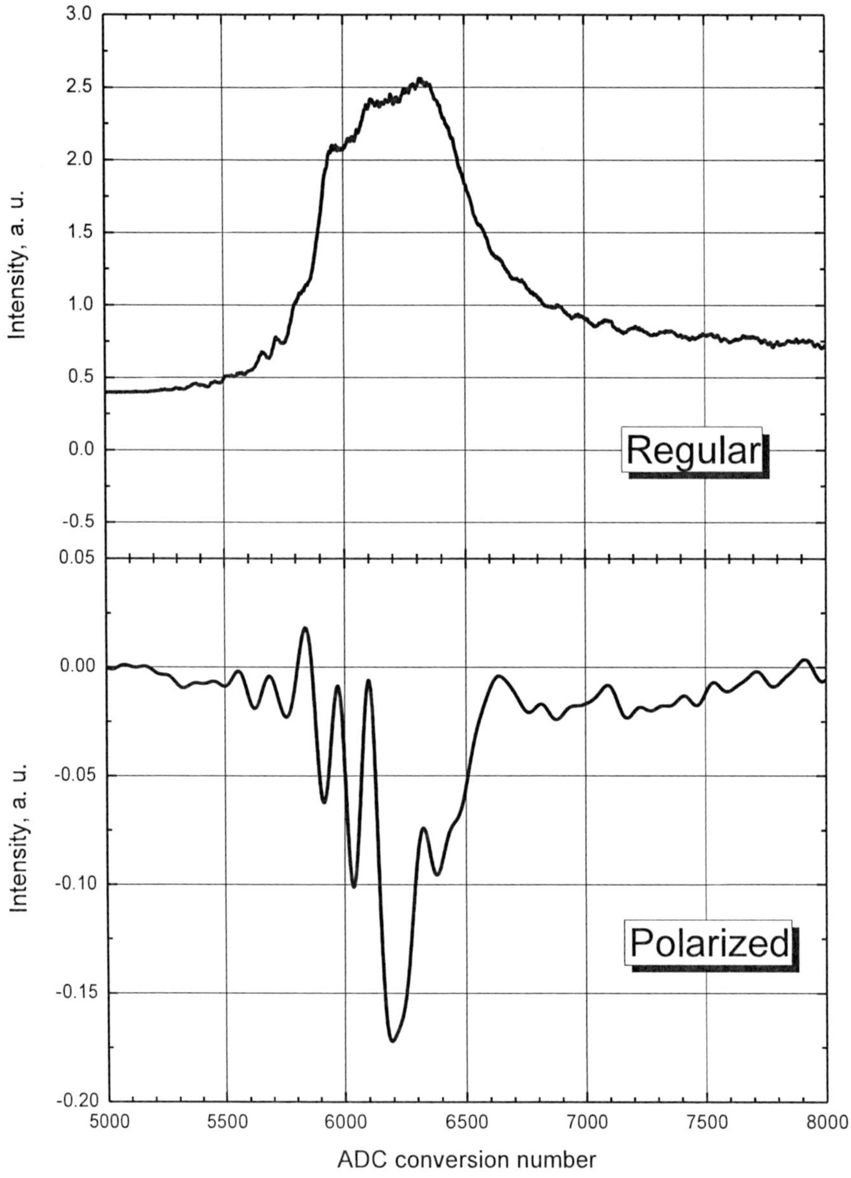

Figure 12. The regular and polarized light-scattering traces of the red blood cell.

The high precision in measurement of cell characteristics helped us to detect the differences in distributions over nucleus volume for the normal lymphocyte sample and for the sample where apoptosis was initiated. The results of this experiment are shown in Figure 11. The lognormal distribution of nucleus volume for the initial sample of lymphocytes (Figure 11 (a)) is characterized by the mean volume of 27.7±0.3 fl, the width of the distribution of 0.227±0.010 and the amplitude of 0.99±0.04. The final distribution (Figure 11 (b)) is formed by normal lymphocytes and lymphocytes with the reduced nucleus volume because of DNA fragmentation in apoptosis [33]. The lognormal distribution of apoptotic cells is characterized by the mean volume of 21.2±0.1 fl, the width of distribution of 0.115±0.004 and the amplitude of 0.48±0.01. The apoptosis was initiated for 51% of cell. This method has a potential to be used in detection of the apoptotic activity of cell population without monoclonal antibodies labeled by fluorescent molecules.

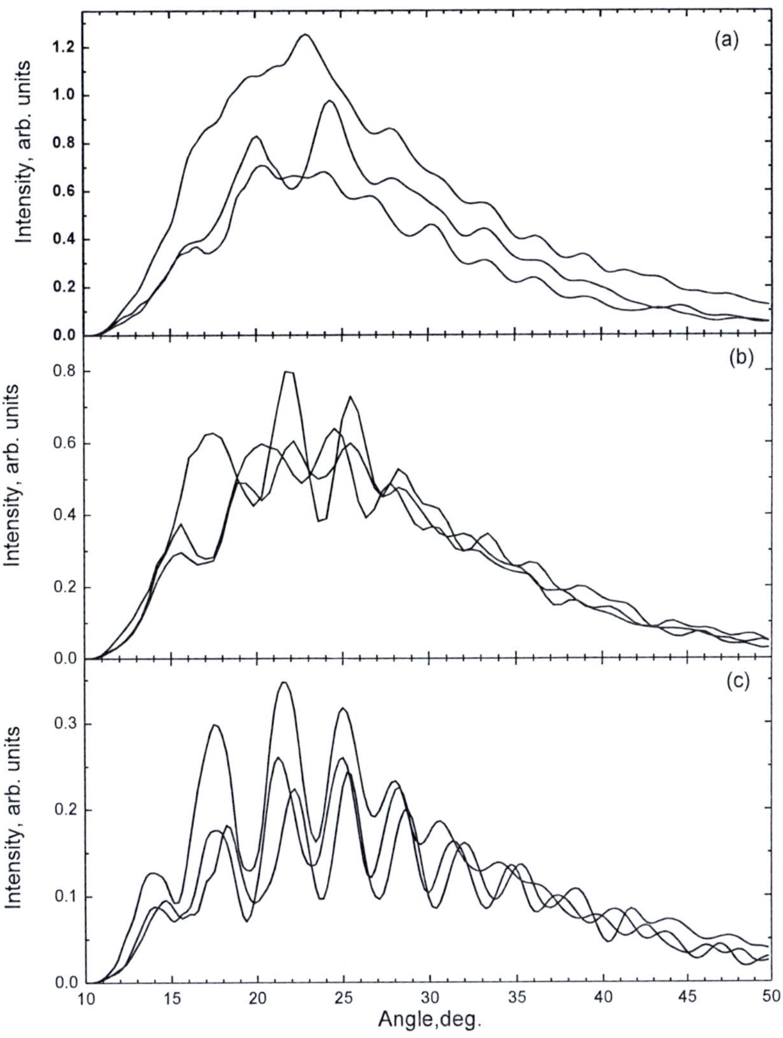

Figure 13. Weighted light-scattering profiles of RBCs in the course of the osmotic hemolysis at different time from the beginning: (a) initial; (b) 1 minutes; (c) 3 minutes.

Red Blood Cells

Unlike a mononuclear cell, the red blood cell (RBC) has a homogeneous internal structure but a quite complex shape under normal conditions. In order to characterize the RBCs the biochemical treatment is used to spherize the RBCs and to solve the ILS problem with Mie theory [9]. The shape of the RBC could be modeled by a biconcave disk. The mathematical formula that describes the shape of an RBC is a field of continuing research [34]. From the instrumental point of view we are ready to support the possible solution of the ILS problem for mature RBCs with maximally available data, i.e. regular and polarized LSPs measured with the SFC (Figure 12). The theoretical simulation of light scattering should be performed with the ADDA program code that means of usage of the nearest-neighbor interpolating within the LSP database for processing of experimental data. We are waiting for an appropriate optical model of a mature RBC to close the characterization problem for these cells. It is desirable that the high precision should be reached in determination of the RBC characteristics in order to be used for the analysis of different processes with these cells like a lysis or shape deformation.

Under normal conditions, a RBC maintains its volume, ionic composition, pH and membrane structure and integrity. When the RBC is placed in a hemolytic medium, it becomes unstable, and its volume changes. If the hemolysis is of the osmotic type, the osmotic gradient is established between intracellular and extracellular media, which causes water to flow into the cell leading to cell swelling. When the volume of the swollen cell exceeds a critical level, the integrity of their membranes is disrupted, allowing the hemoglobin to release into the external media. This is called hemolysis. The lysis observed in a large population of cells follows an S-shaped relation with time, reflecting the time-course of cell volume, the relation between volume and lysis, and the distribution of cell properties in the population. The capacity of RBCs to resist hemolysis can be modified by various pathologies and by physical, chemical or pharmacological treatments. Its study can thus yield relevant information for diagnostic purposes or fundamental researches.

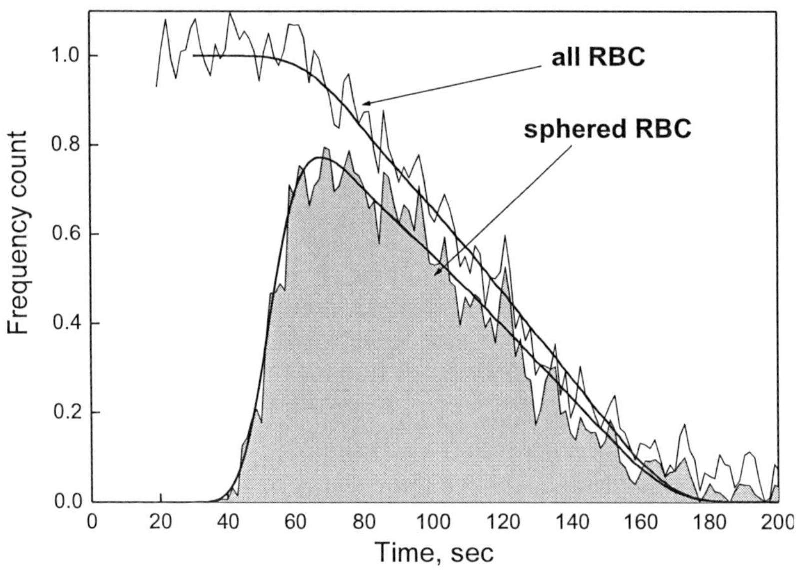

Figure 14. Kinetics of all and sphered RBC recognized from SFC traces.

A complete kinetic model of the hemolysis should contain three main elements: the first is a description of net ion fluxes; the second, cell volume as a function of solute content; and the third, a relationship between cell volume and lysis. There is much information and substantial agreement about many aspects of RBC metabolism, membrane transport, pH control and volume regulation [35, 36, 37, 38, 39]. The non-ideal osmotic behavior of hemoglobin, the change in the net charge on impermeant cell ions with pH, and the kinetics and turnover rates of the main monovalent ion transporters are all well characterized, though in different, often incomparable experimental conditions. Following these models, the physiological properties of erythrocytes can be derived from kinetic volume changes [40]. The last stage of hemolysis, the rupture of the erythrocyte membrane, can be modeled as a spontaneous process by a causal sequence of two thermally activated transitions [41, 42], resulting in the dependence of the membrane breakage rate on the membrane tension. Recently, a complete mathematical model was developed that united all the above mentioned elements [43]. The model was applied successfully for the experimental study of osmotic hemolysis in isotonic solution of ammonium chloride (NH_4Cl) with the SCF technique. SFC was used to recognize sphered RBC from their LSPs (Figure 13) during the process. The obtained kinetics of all and sphered RBC (Figure 14) was compared with that of mathematical model [39] to evaluate the mean permeability and the mean strength (as well as their distributions over the cell population) of the RBC membrane.

Conclusion

We review the state-of-the-art of light-scattering flow cytometry emphasizing the performance in solving the characterization problem for disperse systems using single-particle analysis. We conclude the angle-resolved light-scattering technique provides additional information in flow cytometry to allow us to solve the characterization problem. One possible instrumental solution is the scanning flow cytometry that can be used to measure the regular and polarized light-scattering profiles of individual particles. This technique allows for the determination of four characteristics of individual particles and two angles that describe the particle orientation. At present this is the maximal number of characteristics which can be retrieved with the light-scattering flow cytometry. Summarizing the results described in this chapter we would like to indicate the optical models which were successfully used in the solutions of the ILS problem: a sphere for polymer beads, spherized RDCs and blood microparticles; an oblate spheroid for blood platelets; a hemisphere capped cylinder for rod-like bacterial cells; a sphere with a cover for mononuclear cells. All particles were sized with the sub-diffraction precision. At present the hematology should take into the practice these methods to be available as a routine analysis in clinics in the same manner as RBC volume, RBC hemoglobin, and blood platelet volume.

Flow cytometry is currently the most powerful technology used in the characterization of individual wavelength-sized particles from light scattering. There is no other method that has yet demonstrated statistically proven analyses of disperse systems with sub-diffraction precision in particle sizing. At present the flow cytometry allows one to characterize not only a particle that can be modeled by a sphere, but also particles having rather complex shapes and internal structure. Of course there should be more developments in the characterization of complex particles, but this requires considerable efforts from scientists and especially mathematicians who might not be familiar with the needs of the biologist or life scientist. The life science disciplines have to absorb mathematics as much as possible. Following the lessons of Immanuel Kant, “in any special doctrine of nature there can be only as much proper science as there is mathematics therein” [44]. These efforts should make the life science predictable satisfying the science definition in a short form that is “the development of a prediction procedure which has a calculable precision, even in the presence of perturbations”.

Acknowledgments

We thank our colleagues from the laboratory of Cytometry and Biokinetics (Institute of Chemical Kinetics and Combustion, Novosibirsk, Russia): Konstantin Gilev, Vyacheslav Nekrasov, Maxim Yurkin, Alexander Moskalensky, Darya Orlova, Anastasiya Konokhova.

We also acknowledge financial support by the Russian Foundation for Basic Research (grants 12-04-00737-a, 12-04-31083), by the program of the Russian Government “Research and educational personnel of innovative Russia” (8752 and 8804), and by the Program of the President of the Russian Federation for State Support of the Leading Scientific Schools (grant NSh-65387.2010.4). D. I. Str. also acknowledges the support by the Stipend of the President of Russian Federation for young scientists.

References

[1] Mishchenko, M. I., Hovenier, J. W., Travis, L. D., 1999. *Light Scattering by Nonspherical Particles:* Theory, Measurements, and Applications. Academic Press.

[2] Steinkamp, J. A., Kraemer, P. M. 1979. *Quantitation of lectin binding by cells.* In: Flow Cytometry and Sorting (Melamed, M. R., Lindmo, T. & Mendelsohn, M. L., eds), pp. 397-504. New York: John Wiley & Sons.

[3] Surovtsev,I. V., Razumov,I. A., Nekrasov,V. M., Shvalov, A. N., Soini, J. T., Maltsev, V. P., Petrov, A. K., Loktev, V. B., Chernyshev, A. V., 2000. Mathematical modeling the kinetics of cell distribution in the process of ligand–receptor binding. *J. Theor. Biol.* 206(3), 407–417.

[4] Orlova, D. Y., Borisov, V. I., Kozhevnikov, V. S., Maltsev, V. P., Chernyshev, A. V., 2011. Distribution function approach to the study of the kinetics of IgM antibody binding to FcγRIIIb (CD16b) receptors on neutrophils by flow cytometry. *Journal of Theoretical Biology* 290, 1–6.

[5] Wyatt, P. J., 1968. Differential Light Scattering: a Physical Method for Identifying Living Bacterial Cells. *Appl. Opt.* 7, 1879–1896.

[6] de Grooth, B. G., Terstappen, L. W., Puppels, G. J., Greve, J., 1987. Light-scattering polarization measurements as a new parameter in flow cytometry. *Cytometry* 8, 539–544.

[7] Yurkin, M. A., Semyanov, K. A., Maltsev, V. P., Hoekstra, A. G., 2007. Discrimination of granulocyte subtypes from light scattering: theoretical analysis using a granulated sphere model. *Opt. Express* 15, 16561–16580.

[8] Tycko, D. H., Metz, M. H., Epstein, E. A., Grinbaum, A., 1985. Flow-cytometric light scattering measurement of red blood cell volume and hemoglobin concentration. *Appl. Opt.* 24, 1355–1365.

[9] Semyanov, K. A., Tarasov, P. A., Soini, J. T., Petrov, A. K., Maltsev, V. P., 2000. Calibration-Free Method to Determine the Size and Hemoglobin Concentration of Individual Red Blood Cells from Light Scattering. *Appl. Opt.* 39, 5884–5889.

[10] Rajwa, B., Venkatapathi, M., Ragheb, K., Banada, P. P., Hirleman, E. D., Lary, T., Robinson, J. P., 2008. Automated classification of bacterial particles in flow by multiangle scatter measurement and support vector machine classifier. *Cytometry* Part A 73A, 369–379.

[11] Salzman, G. C., Crowell, J. M., Goad, C. A., Hansen, K. M., Hiebert, R. D., la Bauve, P. M., Martin, J. C., Ingram, M. L., Mullaney, P. F., 1975. A Flow-System Multiangle Light-Scattering Instrument for Cell Characterization. *Clinical Chemistry* 21, 1297–1304.

[12] Loken, M. R., Sweet, R. G., Herzenberg, L. A., 1976. Cell discrimination by multiangle light scattering. *J Histochem Cytochem* 24, 284–291.

[13] Shvalov, A. N., Surovtsev, I. V., Chernyshev, A. V., Soini, J. T., Maltsev, V. P., 1999. Particle classification from light scattering with the scanning flow cytometer. *Cytometry* 37, 215–220.

[14] Dyatlov, G. V., Gilev, K. V., Semyanov, K. A., Maltsev, V. P., 2008. The scanning flow cytometer modified for measurement of two-dimensional light-scattering pattern of individual particles. *Measurement Science and Technology* 19, 015408.

[15] Jacobs, K. M., Lu, J. Q., Hu, X.-H., 2009. Development of a diffraction imaging flow cytometer. *Opt. Lett.* 34, 2985–2987.

[16] Yu, S., Zhang, J., Moran, M. S., Lu, J. Q., Feng, Y., Hu, X.-H., 2012. A novel method of diffraction imaging flow cytometry for sizing microspheres. *Opt. Express* 20, 22245–22251.

[17] Giacomelli, M., Zhu, Y., Lee, J., Wax, A., 2010. Size and shape determination of spheroidal scatterers using two-dimensional angle resolved scattering. *Opt. Express* 18, 14616–14626.

[18] Venkatapathi, M., Rajwa, B., Ragheb, K., Banada, P. P., Lary, T., Robinson, J. P., Hirleman, E. D., 2008. High speed classification of individual bacterial cells using a model-based light scatter system and multivariate statistics. *Appl. Opt.* 47, 678–686.

[19] Bartholdi, M., Salzman, G. C., Hiebert, R. D., Kerker, M., 1980. Differential light scattering photometer for rapid analysis of single particles in flow. *Appl. Opt.* 19, 1573–1581.

[20] Chernyshev, A. V., Prots, V. I., Doroshkin, A. A., Maltsev, V. P., 1995. Measurement of scattering properties of individual particles with a scanning flow cytometer. *Appl. Opt.* 34, 6301–6305.

[21] Yurkin, M. A., Maltsev, V. P., Hoekstra, A. G., 2007. The discrete dipole approximation for simulation of light scattering by particles much larger than the wavelength. *Journal of Quantitative Spectroscopy and Radiative Transfer* 106, 546–557.

[22] Yurkin, M. A., Hoekstra, A. G., 2011. The discrete-dipole-approximation code ADDA: Capabilities and known limitations. *J. Quant. Spectrosc. Radiat. Transf.* 112, 2234-2247.

[23] Orlova, D. Y., Yurkin, M. A., Hoekstra, A. G., Maltsev, V. P., 2008. Light scattering by neutrophils: model, simulation, and experiment. *J. Biomed. Opt* 13, 054057-054057.

[24] Jones, D. R., Perttunen, C. D., Stuckman, B. E., 1993. Lipschitzian optimization without the Lipschitz constant. *Journal of Optimization Theory and Applications* 79, 157–181.

[25] Strokotov, D. I., Yurkin, M. A., Gilev, K. V., van Bockstaele, D. R., Hoekstra, A. G., Rubtsov, N. B., Maltsev, V. P., 2009. Is there a difference between T- and B-lymphocyte morphology? *J Biomed Opt* 14, 064036.

[26] Konokhova, A. I., Yurkin, M. A., Moskalensky, A. E., Chernyshev, A. V., Tsvetovskaya, G. A., Chikova, E. D., Maltsev, V. P., 2012. Light-scattering flow cytometry for identification and characterization of blood microparticles. *J. Biomed. Opt* 17, 057006–1.

[27] Moskalensky, A. E., Yurkin, M. A., Konokhova, A. I., Strokotov, D. I., Nekrasov, V. M., Chernyshev, A. V., Tsvetovskaya, G. A., Chikova, E. D., Maltsev, V. P., 2013. Accurate measurement of volume and shape of resting and activated blood platelets from light scattering. *J. Biomed. Opt* 18, 017001–017001.

[28] Konokhova, A. I., Gelash, A. A., Yurkin, M. A., Chernyshev, A. V., Maltsev, V. P., 2013. High-precision characterization of individual *E. coli* cell morphology by scanning flow cytometry. *Cytometry* Part A 83A, 568–575.

[29] Strokotov, D. I., Moskalensky, A. E., Nekrasov, V. M., Maltsev, V. P., 2011. Polarized light-scattering profile — advanced characterization of nonspherical particles with scanning flow cytometry. *Cytometry* Part A 79A, 570–579.

[30] Maltsev, V. P., 2000. Scanning flow cytometry for individual particle analysis. *Rev. Sci. Instrum.* 71, 243–255.

[31] Mishchenko M. I., Travis L. D., Mackowski D. W. *NASA GISS: Scattering — T-Matrix Codes.* Available at: http://www.giss.nasa.gov/staff/mmishchenko/t_matrix.html. Accessed: June 11, 2013.

[32] Maltsev, V. P., Hoekstra, A. G., Yurkin, M. A., 2011. *Optics of White Blood Cells: Optical Models, Simulations, and Experiments,* in: Tuchin, V. V. (Ed.), Advanced Optical Flow Cytometry. Wiley-VCH Verlag GmbH & Co. KGaA, pp. 63–93.

[33] Maeno, E., Ishizaki, Y., Kanaseki, T., Hazama, A., Okada, Y., 2000. Normotonic cell shrinkage because of disordered volume regulation is an early prerequisite to apoptosis. *Proc Natl Acad Sci U S A* 97, 9487–9492.

[34] Bi, L., Yang, P., 2013. Modeling of light scattering by biconcave and deformed red blood cells with the invariant imbedding T-matrix method. *J Biomed Opt* 18, 55001.

[35] Werner, A., Heinrich, R. 1985. A kinetic model for the interaction of energy methabolism and osmotic states of human erythrocytes. Analysis of the stationary *in vivo* state and of time dependent variations under blood preservation conditions. Biomed. *Biochem. Acta.* 44, 185-212.

[36] Motais, R., Guizouarn, H., Garcia-Romeu, F. 1991. Red cell volume regulation: the pivotal role of ionic strength in controlling swelling-dependent transport systems. *Biochim. Biophys. Acta.* 1075, 169-180.

[37] Joshi, A., Palsson B. O. 1990. Metabolic dynamics in the human red cell. Part III - Metabolic reaction rates. *J Theor. Biol.* 142, 41-68.

[38] Gasbjerg, P. K., Brahm, J. 1991. Kinetics of bicarbonate and chloride transport in human red cell membranes. *J. Gen. Physiol.* 97, 321-349.

[39] Knauf, P. A., Gasbierg, P. K., Brahm, J. 1996. The asymmetry of chloride transport at 38°C in human red blood cell membranes. *J. Gen. Physiol.* 108, 577-589.

[40] Widdas, W. F., Baker, G. F. 1998. The physiological properties of human red cells as derived from kinetics osmotic volume changes. *Cytobios.* 95, 173-210.

[41] Evans, E., Heinrich, V., Ludwig, F., Rawicz, W. 2003. Dynamic tension spectroscopy and strength of biomembrane. *Biophys. J.* 85, 2342-2350.

[42] Hategan, A. P., Law, R., Kahn, S., Discher, D. E. 2003. Adhesively-tensed cell membranes: lysis kinetics and atomic force microscopy probing. *Biophys. J.* 85, 1-14.

[43] Chernyshev A. V., Tarasov P. A., Semianov K. A., Nekrasov V. M., Hoekstra A. G., Maltsev V. P. 2008. Erythrocyte lysis in isotonic solution of ammonium chloride: Theoretical modeling and experimental verification. *Journal of Theoretical Biology* 251, 93-107.

[44] Kant, I., 2004. *Kant: Metaphysical Foundations of Natural Science.* Cambridge University Press.

In: Flow Cytometry
Editor: Stefanos Papandreou

ISBN: 978-1-62808-709-3

Chapter 5

Applications of Flow Cytometry in Environmental Sciences: Inspiring Examples

Piotr Rzymski*[*] *and Barbara Poniedziałek
Department of Biology and Environmental Protection, Poznan University of Medical Sciences, Poznań, Poland

Abstract

Flow cytometry (FCM) is a powerful technology which is routinely used in the medical and veterinary diagnostics. Nowadays it is also an emerging research tool applied in the different fields of environmental studies. This short communication provides the insight to the interesting applications of FCM developed recently in order to study microorganisms, plants and animals. Examples of possible implementation of FCM in environmental monitoring are also given. Major advantages of FCM include automatism and precision of measurements; FCM analyses are time-saving and require small volumes of sample. We believe that an increase in FCM availability on the market and simultaneous decrease of technology costs will firmly establish FCM in the different fields of environmental studies and help to improve already existing methods.

Introduction

Flow cytometry (FCM) is a method to conduct a multiparameter analysis of cells suspended in liquid and passing through a laser beam. As cells pass through the flow chamber they are hit with the laser light beam which is scattered in different directions and recorded as forward light scatter (FSC) and side light scatter (SSC). Simultaneously cells can be excited

[*] Corresponding author: Piotr Rzymski, Department of Biology and Environmental Protection, Poznan University of Medical Sciences, ul. Długa 1/2, 61-848 Poznan, Poland, tel./fax: +48 61 854 9178, e-mail: rzymskipiotr@ump.edu.pl.

with the light of the given wavelengths and the emission of their fluorescence can be measured. However in order to detect any fluorescence, suspended cells require staining with the organic dyes prior to FCM analyses. Some cells, such as these of photosynthetic organisms, contain natural fluorochromes which can be successfully used alone in FCM analyses. At the same time, however, they can hinder the analyses by overlapping the excitation and emission spectra of the synthetic dyes [1, 2].

FCM grows its popularity since late seventies. PubMed search query using MeSH term "flow cytometry" returns 96483 papers published between 1979 and 2012 with over 55% published only in the last 10 years. It is now well-established method used in variety of medical fields including hematology, transplantation, tumor immunology and chemotherapy, prenatal diagnosis and genetics [3-5]. Generally FCM applications concern cell metabolism studies, thus they can be also successfully implemented in selected environmental investigations and contribute with useful results.

Here we aim to briefly present the examples of the inspiring applications of FCM in some recent environmental research. We believe that upcoming years will bring more interesting and useful implementations of FCM in this scientific field.

Microbial Pollution of Drinking Water

It is crucial for global health protection to guarantee the access to safe and standard-meeting drinking water. According to the World Health Organization, the most common and widespread health risk associated with drinking water is microbial contamination [6]. It results from, inter alia, ineffective technologies of sewage and water treatment (or even lack of them) and direct or indirect contamination of surface water with human or animal feces [7]. Drinking such contaminated water or using it for food preparation can lead to the development of waterborne enteric diseases. Pathogenic bacterias that can be present in water and cause human infections include *Escherichia coli*, *Salmonella* spp., *Shigella* sp., *Vibrio cholera*, *Yersinia enterocolitica* and *Campylobacter jejuni* [8]. It is thus necessary to develop effective and precise methods to assess the proper microbial quality of drinking water.

Standard method used to assess drinking water quality has remained unchanged for over 100 years. Bacteria present in water are transformed from the sample into the solid nutrient media, grown at warm temperature and then the colonies are counted. This method is however time consuming (results are available after several days) and owns a great limitation – only mesophilic bacteria capable of reproduction at 20° - 45°C are grown and counted. Therefore the method can only detect a minute fraction of the indigenous bacteria in the water [9]. Other methods used to assess the quality of water are more specific – they are targeted to detect particular bacterial groups or species. For example Colilert-18 test (INDEXX Laboratories) is used for simultaneous detection of both total coliforms and *E. coli* in water with results available in 18 hours [10]. It is thus of general interest to develop and standardize a new method to detect total bacteria in water at different stages of water treatment.

Recently FCM was introduced as a rapid and precise method for bacteria measurements and counts in water [11, 12]. Instead of days it takes only minutes to do necessary analyses. The method is based on staining the cells present in the water sample with fluorescent dyes binding to the bacterial DNA. There are several different dyes that can be successfully applied for such analyses including SYBR Green I (Cyanine), CFDA (Carboxyfluorescein diacetate) or Hoechst [12]. Bacterial cells are then passed through a laser beam in flow cytometer chamber while the resultant scatter and the fluorescence emission of each single cell is recorded by the specific detectors. FCM provides thus an excellent tool to estimate the precise number of total bacteria currently present in the water. The results given by the FCM are far more realistic than those given by the conventional methods. FCM gives an exact count of bacteria in the collected samples instead of necessity to grow bacteria cells for several days under specific conditions which additionally limit and/or eliminate the growth of many bacteria species. FCM can be an excellent tool not only for purely research purposes but because FCM online systems are already available, this method can be applied in monitoring of an entire water supply system and help to identify the problems. Because FCM allows to distinguish between smaller and larger bacterial cells – basing on the DNA content – it can be useful in evaluation of effectiveness of sanitation processes in the water treatment technologies. In December 2012, the Federal Office of Public Health (FOPH) incorporated the flow cytometry method into the Swiss Food Compendium as a recommended test method [13]. Apart from the estimation of the total bacteria count in water samples, the detection of specific bacteria species using FCM has also been recently developed. Füchslin et al. described an interesting application of FCM in detection of pathogenic *Legionella* bacterium. It is consisted of four steps including filtration, double-staining (with FITC- and Alexa-conjugated *Legionella*-specific antibodies), immunomagnetic capturing of cells with anti-FITC immunomagnetobeads, and finally, detection and quantification of cells by FCM. The method allows screening for *Legionella pneumophila* within 3 hours with a detection limit of 500 cells per liter in tap water [13]. Development of FCM coupled with cell-sorting and imaging systems will also contribute in future to the precise identification of specific bacterial species in water.

Apart from the detection of bacteria, FCM was also successfully applied in the detection of other pathogenic microorganisms which can be potentially present in water. FCM has been shown to be a rapid method of *Giardia* cysts and *Cryptosporidum* oocysts identification after staining with immunofluorescent antibodies [14, 15]. It was demonstrated that these stages of protozoa are highly resistant to chlorination – the most common method of water disinfection [16]. Therefore it is necessary to monitor the efficiency of water treatment processes. FCM once again delivers a promising future here as it is fully automated and rapid method that can be potentially applied as an online monitor system and in multi-parametric approach towards assessing biological stability in drinking water distribution networks [12, 17].

Viability of Bees Spermatozoa

Honey bees are a highly valued resource due to their honey and wax production. Globally their population tends to increase although the rate of increase is not keeping pace with actual demands [18]. Moreover, it was shown that in North America and several European countries honey bee population are decreasing. The total colony numbers in Europe decreased from over 21 million in 1970 to about 15.5 million in 2007 [19]. The exact reason behind this situation remains unknown although it is known that the reproduction of bees can be affected by diseases, parasites, use of pesticides and environmental pollution. It was also suggested that these factors may act in combination with each other [18]. Currently (April, 2013) European Commission seriously considers a ban on insecticides in order to improve the situation of honey bees [20]. The fetal toxicity in honey bees can be determined using sperm viability assays. In honey bees *Apis mellifera* L. sperm quality can be evaluated indirectly, by measuring brood quality, i.e., oviposition of fertilized eggs in so-called worker comb cells, egg to larvae proportion, sex ratio of larvae. Direct measures include sperm viability in seminal vesicles (sperm storing organs of males) and ejaculate, as well as in spermathecae (sperm storing organ in females) of queen bees. Several staining methods were tested but the most convenient technique for honey bee sperm viability evaluation was dual staining with SYBR-14 and PI fluorochromes [21-23]. SYBR-14 stains cells green while PI gives the contrasting red color. The particular usefulness of these dyes in honey bee sperm arises from the fact that both of them bind to the nuclear DNA, therefore only the heads of spermatozoon emit fluorescence. This makes single spermatozoon that do not have a pronounced head easily distinguishable and countable under a microscope by a researcher. The other advantage of the dyes is the short time of probe incubation [23]. The estimation of alive:dead spermatozoa ratio is possible by hemocytometer (counting chamber) counts. This method is however time-consuming and imprecise. Hemocytometer counts are subject to the following sources of error non-uniform suspensions, improper filling of chambers, failure to adopt a convention for counting cells in contact with boundary lines or each other, and statistical error. The other method is fluorometric quantification of honey bee sperm [23, 24].

Fluorchromes used in dual-staining of honey bee sperm allow the application of FCM. Moreover, FCM allows the fluorescence emission to be measured in a rapid automated way. FCM has already been shown to be an excellent method to quickly determine sperm numbers in ant queens [25]. Recently, two independent research teams from Poland demonstrated that the viability of honey bee semen can be estimated by staining sperm with DNA-binding fluorochromes followed by FCM estimation of dead:alive ratio [26, 27]. Maximum fluorescence emission for propidium iodine is 617 nm while for SYBR-14 - 516 nm. Emission spectra of these fluorochromes does not overlap and can be thus recorded separately in the different channels. Using a dot-plot with PI vs. SYBR-14 emission, two separate populations of semen can be observed. Cells stained only with PI are treated as dead while cells with definitely higher fluorescence in the red fluorescence channel are considered as alive. The only unclear issue is that part of the living population can also emit low fluorescence in the PI channel. Dual fluorescence (green and red) has previously been reported in honey bee sperm studies based on the microscope analyses in which it constituted up to 0.13% of the sperm population or was described as a 'small subpopulation' [23, 28]. This phenomenon probably results from the injuries to spermatozoa in the staining solution. In studies of the different insect taxa, double-stained sperm was often considered as viable and possessing fertilizing abilities [29-31].

FCM demonstrates usefulness in studying various aspects of honey bee reproduction. Furthermore, given that SYBR-14 and PI stains have been successfully used for other insects, it can be argued that this method could be used in ecological research to test the impact of environmental (physicochemical and biological) factors on honey bees (and other insects) spermatozoa, as well as in evolutionary research.

Monitoring the Cyanobacteria

Cyanobacteria are a common component of the phytoplankton communities and contribute to the primary production in aquatic environments. Apart from their important ecological role, development of cyanobacteria has also been associated with potential health threats. Under optimal conditions such as high concentration of nutrients, windless and sunny weather and pH >6, cyanobacteria can form massive dense blooms [32, 33]. Several bloom-forming species have been identified to produce secondary metabolites of uncertain biological role although inducing the toxic effects in humans and animals The exposure can lead to hepato-, dermato- or neurotoxicity and in extreme cases can - death. Toxic species of cyanobacteria include, among many: *Microcystis aeruginosa*, *Cylindrospermopsis raciborskii*, *Planktothrix rubescens*, *Aphanizomenon flos-aquae* and *Lyngbya majuscula* [34, 35]. Cyanobacterial blooms are observed worldwide (in both marine and freshwater environment) and their number appears to be growing due to the degradation of surface water (mainly via cultural eutrophication) [36]. Thus there is a need to develop an effective monitoring and forecasting tools for public health protection. Probably the most sophisticated although expensive method was introduced in 2011 for Lake Erie (USA). It is based on images taken by European Space Agency (MERIS) satellite system. People can sign up for the emailed bulletin that was developed at the Center of Excellence for Great Lakes in Human Health at the National Oceanic and Atmospheric Administration and warns swimmers before they decide to visit a lake [37]. In other countries less expensive but time-consuming methods based on taxonomical identification, analyses of chlorophyll-a concentration and physico-chemical conditions of water are being applied in forecasting a possibility of cyanobacterial blooms during recreational season. It is very important to develop an automatic monitoring system particularly for these public bathing areas which are open to a large number of people and at the same time are known to be potentially threatened with toxic cyanobacterial species.

FCM has already been applied in studies on the phytoplankton [38]. Cells of these aquatic organisms possess photosynthetic pigments including chlorophylls which can be excited in FCM and then their fluorescence can be recorded. It was demonstrated that fluorescence excitation and emission can be successfully applied in the classification of groups such as cryptophytes, chlorophytes, bacillariophytes and cyanobacteria based upon spectrally similar accessory pigments within these groups [38, 39].

Apart from the red algae (Rhodophyta), cyanobacteria are the only organisms to possess phycobilins as an accessory photosynthetic pigments with phycocyanin (PC) being the most commonly found in their cells. The excitation spectra of PC does not overlap chlorophyll-a and its maximum ranges between 615 and 620 nm while its maximum emission can be found at ~650 nm [40]. Several studies demonstrated a positive correlation between PC fluorescence and cyanobacterial biomass during *Microcystis aeruginosa* blooms [41, 42]. However PC fluorescence is not a direct indicator of toxins concentration in water and species identification using this parameter is also not possible, it was found that PC fluorescence can be applied as a simple and sensitive indicator of cyanobacteria in water and serve as a tool that can provide an early warning of the presence of potentially toxic cyanobacterial metabolites in water. So far only fluorometers were applied in the monitoring of the PC fluorescence [41-43]. FCM is however more sensitive and precise method and could be potentially applied in online systems of cyanobacteria monitoring in surface water.

Furthemore, FCM could be also applied in order to monitor the presence of cyanobacterial cells in drinking water and evaluate the efficiency of water treatment systems.

Conclusion

The advantages of FCM include accurate counting and simultaneous measurement of cell properties (size, volume and granularity) as well as precise recording of the fluorescence emission (labeled or natural). Furthermore, FCM is a rapid, automatic and reproducible technique. The decrease of technology costs and increase in equipment distributors positively popularize FCM. It can be expected that novel and interesting methods using FCM will be developed while the existing FCM assays will be improved. It is beyond any doubt that FCM can be useful in the different fields of environmental science and as it can be drawn from the given examples this method can find its way in both basic and applied research.

References

[1] Nunez, R. (2001). Flow Cytometry: Principles and Instrumentation. *Curr. Issues Mol. Biol.* 3, 39-45.

[2] Macey, M.G. (2007). *Flow Cytometry: Principles and Applications*. London, Humana Press.

[3] Dieterlen, M.T., Eberhardt, K., Tarnok, A., Bittner, H.B., and Barten, M.J. (2011). Flow cytometry-based pharmacodynamic monitoring after organ transplantation. *Methods Cell Biol.* 103, 267-284.

[4] Kaczmarek, A., Mackiewicz, A., Leporowska, E., and Osawa, T. (2002). The role of flow cytometry in clinical diagnosis. *Contempor. Oncol.* 6, 366-373.

[5] Wedemeyer, N., and Pötter T. (2001). Flow cytometry: an 'old' tool for novel applications in medical genetics. *Clin. Genet.* 60, 1-8.

[6] WHO (2000). *Global Water Supply and Sanitation Assessment*. Geneva, World Health Organization.

[7] Fawell, J., and Nieuwenhuijsen M.J. (2002). Contaminants in drinking water. Environmental pollution and health. *Br. Med. Bull.* 68, 199-208.

[8] Dupont, H.L. (2013). *Infections of the gastrointestinal tract: microbiology, pathophysiology, and clinical features*. London,Springer.

[9] Hammes, F., Berney, M., Wang, Y., Vital, M., Köster, O., and Egli, T. (2008). Flow-cytometric total bacterial cell counts as a descriptive microbiological parameter for drinking water treatment processes. *Water Res*. 42, 269-277.

[10] Warden, P.S., DeSarno, M.S., Volk, S.E., and Eldred, B.J. (2011). Evaluation of Colilert-18 for detection and enumeration of fecal coliform bacteria in wastewater using the U.S. Environmental Protection Agency Alternative Test procedure Protocol. *JAOAC Int.* 94, 1573-1580.

[11] Hammes, F., and Egli T. (2010). Cytometric methods for measuring bacteria in water: advantages, pitfalls and applications. *Anal. Bioanal. Chem.* 397, 1083-1095.

[12] Drinking water unexpectedly rich in microbial life. *Science Daily*. Retrieved April 21, 2013.

[13] Füchslin, H.P., Kötzsch, S., Keserue, H.A., and Egli, T.(2010). Rapid and quantitative detection of Legionella pneumophila applying immunomagnetic separation and flow cytometry. *Cytometry A* 77, 264-274.

[14] Ferrari, B.C., and Veal, D. (2003). Analysis-only detection of Giardia by combining immunomagnetic separation and two-color flow cytometry. *Cytometry A* 51, 79-86.

[15] Hsu, B.M., Wu, N.M., Jang, H.D., Shih, F.C., Wan, M.T., and Kung, C.M. (2005). Using the flow cytometry to quantify the Giardia cysts and Cryptosporidium oocysts in water samples. *Environ. Monit. Assess* 104, 155-162.

[16] Betancourt W.Q., and Rose, J.B. (2004). Drinking water treatment processes for removal of Cryptosporidium and Giardia. *Vet. Parasitol.* 126, 219-234.

[17] Lautenschlager, K., Hwang, C., Liu, W.T., Boon, N., Köster, O., Vrouwenvelder, H., Egli, T., and Hammes, F. (2013). A microbiology-based multi-parametric approach towards assessing biological stability in drinking water distribution networks. *Water Res.* 47, 3015-3025.

[18] Vanengelsdorp, D., and Meixner, M.D. (2010). A historical review of managed honey bee populations in Europe and the United States and the factors that may affect them. *J. Invertebr. Pathol.* 103 Suppl 1, 80-95.

[19] FAO (2009). FAOSTAT. *Food and Agriculture Organization of the United Nations* (FAO).

[20] EU could impose pesticide ban to protect bees. Reuters. Retrieved April 21, 2013.

[21] Locke, S.J., Peng, Y-S., and Cross, N.L. (1990). A supravital staining technique for honey bee spermatozoa. *Physiol. Entomol.* 15, 187–192.

[22] Peng, Y-S., Locke, S.J., Nasr, M.E., Liu, T.P., and Monatgue, M.A. (1992). Differential staining for live and dead sperm of honey bees. *Physiol. Entomol.* 17, 269 –276.

[23] Collins, A.M., and Donoghue, A.M. (1999). Viability assessment of honey bee, Apis mellifera, sperm using dual fluorescent staining. *Theriogenology* 51, 1513–1523.

[24] Reichardt, A.K., and Wheeler, D.E. (1995). Estimation of sperm numbers in insects by fluorometry. *Insects Soc.* 42, 440 –452.

[25] Cournault, L., and Aron, S. (2008). Rapid determination of sperm number in ant queens by flow cytometry. *Insects Soc.* 55, 283–287.

[26] Rzymski, P., Langowska, A., Fliszkiewicz, M., Poniedziałek, B., Karczewski, J., Wiktorowicz, K. (2012). Flow cytometry as an estimation tool for honey bee sperm viability. *Theriogenology* 77, 1642-1647.

[27] Tofilski, A., Chuda-Mickiewicz, B., Czekońska, K.,Chorbiński, P. (2012). Flow cytometry evidence about sperm competition in honey bee (Apis mellifera). *Apidologie* 43, 63-70.

[28] Den Boer, S.P., Boomsma, J.J., and Baer, B. (2009). Honey bee males and queens use glandular secretions to enhance sperm viability before and after storage. *J. Insect. Physiol.* 55, 538–543.

[29] Damiens, D., Bressac, C., Brillard, J.P., and Chevrier, C. (2002). Qualitative aspects of sperm stock in males and females from Eupelmus orientalis and Dinarmus basalis (Hymenoptera: Chalcidoidea) as revealed by dual fluorescence. *Physiol. Entomol.* 27, 97–102.

[30] Hayashi, F., Tsuchiya, K. (2005). Functional association between female sperm storage organs and male sperm removal organs in calopterygid damselflies. *Entomol. Sci.* 8, 245–252.

[31] Nakahara M, and Tsubaki Y. (2007). Function of multiple sperm-storage organs in female damselflies (Ischnura senegalensis): difference in amount of ejaculate stored, sperm loss, and priority in fertilization. *J. Insect. Physiol.* 53, 1046 –1054.

[32] Mur, L.R., Skulberg, O.M. and Utkilen, H. (1999). Cyanobacteria in the environment. In I. Chorus, and J. Bartram (Eds.), *Toxic Cyanobacteria in water: a guide to their public health consequences, monitoring and management* (14-40). Spon, London.

[33] Havens, K.E. (2008). Cyanobacteria blooms: effects on aquatic ecosystems. *Adv. Exp. Med. Biol.*, 619, 733-747.

[34] Rzymski, P., Poniedziałek, B., and Karczewski, J. (2011). Gastroenteritis and liver carcinogenesis induced by cyanobacterial toxins. *Gastroenterol. Pol.* 18, 159-162.

[35] Rzymski, P., Poniedziałek B., (2012). Dermatotoxins synthethised by blue-green alga (Cyanobacteria). *Post Dermatol. Alergol.,* 29, 47-50.

[36] Vasconcelos, V. (2006). Eutrophication, toxic cyanobacteria and cyanotoxins: when ecosystems cry for help. *Limnetica*, 25, 425-432.

[37] Stumpf, R.P., Wynne, T.T., Baker, D.B., and Fahnenstiel, G.L. (2012). Interannual variability of cyanobacterial blooms in Lake Erie. *PLoS One* 8: e42444.

[38] Dubelaar, G.B.J., and Jonker, R.R. (2000). Flow cytometry as a tool for the study of phytoplankton. *Sci. Mar.* 64, 135-156.

[39] Hofstraat, J. W., de Vreeze, M.E.J., van Zeijl, W.J.M., Peperzak, L., Peeters, J.C.H., Balfoort, H.W. (1991). Flow cytometric discrimination of phytoplankton classes by fluorescence emission and excitation properties. *J. Fluoresc.* 4, 249-265.

[40] Pizarro, S.A., and Sauer, K. (2001). Spectroscopic study of the light-harvesting protein C-phycocyanin associated with colorless linker peptides. *Photochem. Photobiol.* 73, 556-563.

[41] Izydorczyk, K., Tarczynska, M., Jurczak, T., Mrowczynski, J., and Zalewski, M. (2005). Measurement of phycocyanin fluorescence as an online early warning system for cyanobacteria in reservoir intake water. *Environ. Toxicol.* 20, 425-430.

[42] McQuaid, N., Zamyadi, A., Prévost, M., Bird, D.F., Dorner, S. (2011). Use of in vivo phycocyanin fluorescence to monitor potential microcystin-producing cyanobacterial biovolume in a drinking water source. *J. Environ. Monit.* 13, 455-453.

[43] Zamyadi, A., McQuaid, N., Prévost, M., and Dorner, S. (2012). Monitoring of potentially toxic cyanobacteria using an online multi-probe in drinking water sources. *J. Environ. Monit.* 14, 579-588.

Index

A

B

C

D

E

F

G

H

I

J

K

L

M

N

U

V

W

X

Y